A Guide to Un
Birth Chart, Sto
Relationship Partner

ASTROLOGY FOR *Beginners*

As Written in the Stars

ASHLEY BRITTANY

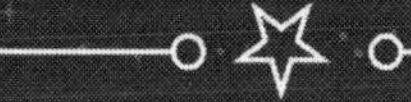

Astrology for Beginners: A Guide to Understanding Your Birth Chart, Star Sign, and Ideal Relationship Partner

---

As Written in the Stars

Ashley Brittany

Published by *Master Plan Focus.*

Contents

To my publisher, Andy,

Thank you for believing in me and for giving me the platform to be able to express myself.

To my assistants, Bex and Kate,

Thank you for supporting me along this journey - it was hard, but look how far we have come!

To my dear friend, Chelsea,

Thank you for being a beacon of knowledge in my astrological awakening!

And to all astrology fans across the world,

I hope this book serves you well! x

We can guide you through this book!

Download our free journal today!

This Astrology Journal will guide you through the book and allow you to record your findings about yourself as you go! By the end of the book you should have a good understanding of your big three signs and a much greater insight into your *true* self!

Want to know how to read your Birth Chart?

Download the guide now and learn to read your chart in just 5 minutes!

Introduction

Have you ever felt like there is more to life than what you are living? Perhaps you have felt that there are some changes that you need to make but you don't know where to begin. If you have been dreaming of a different life, then I have a little secret to tell you... there is a guide, or some may call it a map, of the ways that you can live your very best life. This personal guide was written at the moment you were born, and it is a path that you can study and follow to help you make decisions that will align you with your dream life.

This guide is called 'astrology' and it is a practice that has been studied and refined over thousands of years. It is a tool that can explain your personality in a way that only you feel seen and heard. It is an expression of your life, written in the form of planets, zodiac signs, houses and conjunctions. As you learn the language of astrology, you can begin to read your own birth chart and find alignment in your life.

Using astrology is a way to remove yourself from the rat race that you may find yourself in. It is a way to tune into your inner self and to find the purpose that you have been seeking. If you are ready to discover these secrets, then you are in the right place.

From ancient cultures to modern billionaires, astrology has been used to help find the path to success. Local shamans of tribal communities would use the stars to find out information and purpose about new babies as they were born. Even today, this ancient practice is used in many and varied ways. The wealthy billionaire J.P. Morgan used astrologists to help him to choose the right time and place to make his investments, and he sure did see the positive return!

Now it is common practice for people who are seeking purpose to consult astrologists to find their aligned calling. If you are wondering if this could also be for you, the answer is yes! Astrology is for everyone. Let me tell you about a client of mine...

A few years ago, I had a client come to see me for an astrological reading – specifically a birth chart consultation. She was 29 years old and working hard, yet her bank account was always falling to zero at the end of each month. She was desperately seeking guidance, not only financially, but also to find a loving and supportive partner, and to seek satisfaction in her daily life. She had a feeling that she could make some changes to achieve her dream life, yet she had no idea where to start. After spending one hour together, diving into the meaning of her birth chart, she left my office in tears. She was so grateful to have rediscovered the truest version of herself. All she needed was to remember her strengths and find the tools to overcome her challenges. With astrology, she was able to recall her earliest memories of activities and ways of living that really did light up her heart. She had found who she was looking for, and that was herself.

After six months, I ran into this same client in the supermarket. With recognition in her eyes, she came up to me, radiating a bright and vibrant energy from every part of herself! She told me about her fiancé, the house she had just put a deposit down for and the workshops she had started hosting to bring women together. She was abundant in every aspect of her life and it was a true pleasure to see this powerful transformation.

This client of mine is a living example of just how much can change simply by looking to the stars for guidance. Do some of the aspects of this client's life resonate with your own?

If so, there is no need to be ashamed, for in this fast paced 21st century world we are constantly bombarded with opportunities that are available to us at the click of a button, or a simple internet search. Although this wonderful technology has made communication easier than ever, it is becoming more and more difficult to know who we truly are. Understanding our own self and our own path is a practice that takes time and patience to understand. Looking to astrology for the answers is often like finding the key that has been lost for 10 or even 20 years, and now being able to unlock and remember the treasures that you had hidden away in the chest of your heart.

Now by understanding astrology, it doesn't mean that life won't have its challenges anymore, for there will always be the darker moments that occur during your life path. But the good news is that with astrology you can be better prepared for these moments of hardship, and therefore create the right mental and physical stamina to emerge from these events with a positive attitude.

It has taken me nearly 20 years to accumulate the knowledge in this book, and the most wonderful part about this time frame is that I have seen so many personal transformations. I often tell a story to my clients about how the art of astrology is similar to the lifecycle of a butterfly.

We are born at a specific time and place, just like that of a caterpillar. Upon birth we really have no idea about our full potential as a human being. We go to school, work, in and out of relationships, and before you know it, you have forgotten about the significance of being a special soul on this earth at this time. When the time comes to look back to your birth chart you enter the stage of the chrysalis. It can take a bit of time here as you begin to uncover what the planets mean for you, or to rediscover how your emotional ways of being are expressed in the world. From social media to fast and frequent news, we are so often trying to keep up with others that we forget to keep up with ourselves. When the time comes to reconnect with your true self, the rebirth begins. Time and space slow down as you work through the aspects of your chart that present challenges, and those that create flow and ease. This is a time of reading, comparing, practicing and trying new things as you prepare to emerge from your cocoon. When you have found the essence of yourself that you are looking for you know that it can never be taken away from you, and so the transformation happens.

Overnight, you become the beautiful, vibrant butterfly that is ready to fly. The one that is unafraid to be seen as it's full self, for you know that your weaknesses can never be used against you. You are ready to shine and radiate your beauty to those around you, in the hope that they too find the courage and time to look at their own birth charts and to find the way in which you can fly together.

If you are ready for these insights or to make a change in your life , you can look to a well-studied astrologist for help. Astrologists are able to look at your chart and explain your personality in ways that you may have never heard of. We are able to predict the challenging moments in your life, but also the happiest and most abundant times too. We can also look at the chart of your partner and tell you how your relationship will play out over time. If you are seeking a partner, we can help you to find the right one by looking at what it is that you need in a relationship! We are able to get you back into alignment with the parts of yourself that you have lost over time. Working with an astrologist can truly make your life easier and can make life transitions a lot smoother.

Can you imagine quitting your current job and finding one that you actually love? Can you visualize yourself waking up each morning and feeling excited for the day ahead because you are passionate about your own life? Just imagine what it would be like to have financial abundance yet also be living your soul's purpose. To feel valued in your work and by your friends and family. If you really want to be seen and heard in all parts of your life, then it is time to see an astrologer. I have helped so many people to create a life that feels like freedom, simply by understanding this ancient practice of astrology.

Another client of mine, left a wonderful testimonial for the changes that she experienced after working with me.

"Ashley helped me to let go of the boring office job I was stuck in for 36 years. I'm now 51 years old and living the life of my dream in Bali, doing yoga every day, creating events where people come together to share feminine energy, wisdom, and love."

That is why I have created this book, to encourage you to start your own journey into alignment.

As you work through the chapters in this book you will understand the origins and use of astrology since the beginning. You will get to know the 12 zodiac signs for what they truly are (and let me tell you – there is a lot more to know than just your daily horoscope in the newspaper or magazine column). You will learn about the elements of the signs, and what that means for your own personality. Plus, you will begin to decipher the meaning of the planets and how that can highlight your biggest lessons and blessings in this lifetime. By the end of this book, you will be able to look at your very own birth chart and understand the placement of your planets in each of the houses. You will have a great overview of how the look of your chart creates your personality and qualities. You will even know how to search for the perfect partner, one that will be supportive and encouraging of all your quirky characteristics!

So many people that turn to astrology have never looked back. This is wisdom that can help in all aspects of your life.

> *"I was always judging my weaknesses. Ashley taught me that every weakness can be my strength if I allow it to be."*
>
> *"Working with Ashley has totally changed my love life. Finally, I'm meeting men with whom I feel aligned and that really do understand me. Dating has become easier than ever before."*

I would not be teaching this work if I didn't fully believe in it myself. Seeing people gain the courage to make changes in their life is one of the most powerful acts to witness. If I can give you the simplest reminder of one aspect of your chart that will alter your life's trajectory, then I will not hold back. Now is a time to be authentically you. More so than ever before. We do not need any more of the same people in this world, we need the freedom seekers, the ones who take on self-development tasks and enjoy the challenge. We need the ones that are accepting themselves fully and unconditionally. This book will give you these tools and so much more. Not only will you understand yourself better, but you will also be feeling more energized and joyful in your life as you move closer into alignment with your true self.

Dare to be different and dare to be you. The stars have got a plan for you, and your life will become far more pleasant as you work with them and not against them.

My only question for you is, are you ready?

CHAPTER 1

UNDERSTANDING ASTROLOGY

WHAT IS ASTROLOGY?

Astrology is the study of correlations between celestial and earthly events.

It looks at the position of the planets in the sky and considers how these placements affect human life and events in the natural world.

The stars and planets in our universe have been studied for thousands of years and by many different cultures which has given us the opportunity to bring all of this wisdom together to refine the practice of astrology.

There is great knowledge within the practice of astrology, and as you journey through this book, you too will begin to feel the influence of the stars more strongly. In regard to science, astrology has no official evidence or large-scale studies. That is because each person is unique and to study one's birth chart thoroughly can take many years, or even a lifetime!

To create a study would mean that the participants would have to actually live through all of the major planetary transitions and then reflect back on what was happening in their life at that point in time.

Another main reason that the evidence of astrology has not been well documented, is because there is a high level of personal interpretation.

The accuracy of your reading can very much depend on the astrologist that you consult, and this is why it is important to be able to read your own chart, so that you can eliminate personal bias!

But the good news is, that you don't need scientific studies to prove that astrology works, you simply need to understand your own astrological aspects and how they create your life. With time spent in reflection, you will be a living testimonial to the art of astrology and you will see for yourself how this language of the planets has the ability to influence your own life.

The American Radio Engineer John Nelson's discovery in 1951[1] is an important event to note. He observed that the position of the planets had an effect on the degradation of radio wave signals (1). This was one of the first, measurable changes that showed how the planets affected life on earth.

Aside from this, the proof lies in the historic use of astrology. From ancient civilizations and shamans to modern horoscope technology, this is an art that has travelled the world far and wide and been used by many cultures to gain accurate insights into human life. We will study this further throughout the chapter.

Does Astrology really work?

Now, if you ask someone if astrology really does work, you are sure to receive a different answer from every person. It is best to do your own research and to apply the lessons in this book to your own birth chart and see if the evidence stacks up to be true! *Here's a little hint: I wouldn't have spent the time putting this book together if I didn't believe in it!*

Once you understand the basics, you will gain so much more insight and information out of the situations that are unfolding in your life and the world around you. This is why I truly believe that astrology works for everyone!

Why believe in Astrology?

Once you understand the practice of astrology, you can use this information to relate to your family and friends differently. You will gain clarity about what it is that you seek in a partner. You will be aware of the challenging situations that may arise in your future, which will give you the time and space to prepare for these events unfolding in your life. Astrology offers comfort and trust in the future. It gives you a glimpse of the greater meaning of life. When you become confident in understanding how astrology works, you can also use it in relation to those around you to understand their personality better. This will help to strengthen your relationships and hopefully reduce conflict in your life.

For thousands of years, ancient tribes have looked to the sky for messages, and omens. They had an innate knowledge that the sky was talking to them and they felt the powerful connection of the macrocosm and microcosm. We want to continue to look to the sky for the next thousand years to continue to understand the wisdom that our elders spoke about.

We all want to be sure about our futures, to reduce conflict with others, and understand that things happen for a reason. Through astrology, I believe that greater world peace can be created, because if everyone looked inwards, just for a moment or two, we would realize that we are all made up of the same stardust. We are all under the influence of the greater celestial bodies and we are all connected to each other, the earth and the sky.

The History of Astrology

Even before astrology was used as a method to understand life, there were indigenous cultures around the whole world who looked to the sky for omens. They would interpret messages from the skies, the moons and the stars and share these messages with one another. The stars would tell them when to move on to a new location, when the weather will turn, how the crops will grow, and which plants are available for foraging. The stars were a crucial part of everyday life, for they guided the actions and wisdom of the ancient tribes. Archaeologists have found mammoth tusks inscribed with the phases of the moon and cave paintings with maps of the stars in the sky. In fact, the stars were mapped long before the earth was and people had a better understanding about what was happening in the sky than the mysteries that existed on the other side of the planet.

Thanks to all of the astronomers and astrologists that have come before us, including the likes of Galileo Galilei, Paracelsus, Nicholas Culpeper, Hipparchus,

Nostradamus, Ptolemy, and many more, the studies of the skies were documented and passed onto others.

There have been many ground-breaking discoveries about planets, stars, the way the earth moved, and the constellations. These discoveries have been made throughout many cultures. Often, the accuracy of these discoveries improved as the information reached more and more people. This long history of looking to the stars is has now evolved into what is known as 'Western Astrology'. Western Astrology is the study of the planets, houses, stars and their movement and it is a way of thinking that dates back to the oldest civilizations.

The first known use of astrology originates in Mesopotamia around 3000 BC. It was here that the Babylonians discovered that certain planets, or as they were known back then, 'wandering stars', were moving through space in the sky. When this space was divided into 12 segments (now known as the zodiac belt), astrologists were able to track the planets and record time. The 12 sections of the zodiac belt are split up into what we now commonly know as the 12 zodiac signs – Aries, Taurus, Gemini, Cancer, Leo, Virgo, Libra, Scorpio, Sagittarius, Capricorn, Aquarius and Pisces. These were recognized by the unique star constellations in the sky that represented each of the animals or symbols.

From here, astrology began to spread over Europe very quickly. In the 6th Century BC, the Ancient Greeks started using the planets as their guides and as their gods. The Greeks made significant advances in both astrology and astronomy. They began to personify the planets into godlike creatures, making astrology more dramatic and exciting. They would tell elaborate stories about how each sign came about and why they hold certain energies. These creative tales are often retold today as a fun way to remember each of the signs.

Moving to the continent of India, around 1000 BC, astrology was very popular, and still is to this day. The Indian people had their own style of astrology, known as Vedic Astrology, which differs greatly from the most common western astrology that is used today. As Vedic astrology, also known as Indian moon astrology, is still practiced today too, we will highlight the major differences between these styles of astrology later in this chapter.

It didn't take long until astrology was adopted in East Asia too. In ancient China, it was very common for new-born babies to have a horoscope cast for them at their time of birth. Like the Indian people, the people of China also created their own unique adaptation of Greek astrology. Their style of astrology uses 12 animals, as figureheads, and moves in a 12-year cycle- however, the animals vary greatly from the western astrology signs. For example, you may have heard of the year of the Rat or the year of the Dragon. They also work a lot with the elements, but rather than the 4 Western elements (earth, air, fire and water), they use 5 Chinese elements (fire, wood, metal, earth and water).

Although India and China both took on their own astrological styles, the western world continued to use the beliefs from the Ancient Greeks.

As humanity progressed into the Middle Ages, astrology was a common part of the culture. Doctors, mathematicians, and astronomers all used the placement of the stars and planets in the sky to make diagnoses and to find the answer to complex formulas. Astrology was an accepted and widely-used practice at this time. It provided comfort in times of uncertainty and a path forward to search for purpose and meaning in many peoples' lives.

It was only when the Church gained more power, that the use of astrology began to dwindle. Anything that was deemed to be more powerful than God was simply not spoken about or studied, and so talk of the planets and their powers decreased significantly.

Perhaps the most famous example of this is Galileo Galilei the Italian astronomer, physicist and engineer. Galileo was almost murdered because of his astrological beliefs and faced trial and conviction for his theory that the earth rotated around the sun. He was shunned and resultantly, the public were too afraid to speak about his theories in case they too were convicted of challenging the Church.

Next came the Age of Enlightenment in which scientific discovery flourished. This created another negative turn for astrology, which was seen as an ancient way of thinking rather than a modern, evidence-based practice. With the search for new and better theories, astrology was almost completely neglected. Luckily, during these times, there were many underground astrologists, herbalists and seers that kept the ancient knowledge and wisdom alive. It was from their books and written knowledge that we are able to uncover their age-old discoveries and rekindle the lost art of astrology.

In this day and age, the use of astrology is becoming more widespread than ever. Knowledge is re-emerging as many people start to look to our ancestors and older civilizations to restore the wisdom that was once commonly known. This innate search is a natural part of us, as we all long for knowing the truth of who we are and where we came from. Hence, the study of astrology, and the enlightenment it can evoke, can be applied so diversely to modern life.

ASTROLOGY AND TECHNOLOGY

Thanks to modern day technology, a simple internet search can bring up thousands of years of astrological wisdom. Before the internet, you would have had to consult a trusted astrologer to receive a map of your birth chart. But these days, you can simply input your birth date, time and place into a 'birth chart calculator' and have a copy of your chart within seconds! Terms such as 'sun signs', 'horoscopes' and 'astrology' have reached an all-time high on internet search engines as more people are intrigued to find out their personality strengths. There are even apps that will provide daily affirmations for your sun sign, or daily horoscopes and overviews of the week ahead, depending on your unique planet placement.

Technology can be a blessing in many ways, as it allows us to communicate quickly and efficiently over the invisible internet. (Unless, of course, Mercury is in Retrograde and communication chaos is everywhere!) This is also positive in a sense that many people have access to astrological insights about their life and how to make positive changes according to their planetary placement.

The negative side of technology is that often there can be conflicting information found on the internet. Many sites provide generic information about one's sun sign, but they forget to show the importance of looking at the whole chart. There are also many scams and internet 'gurus' who have no background in astrology. But the key to avoiding this misinformation is learning your own chart well, and applying your own basic astrological knowledge to your life. If something intuitively feels right, then trust it. But if you feel like someone is trying to sell you a service or misuse your personal information, then question it!

Just like listening to your intuition and learning to trust yourself, listening to what you hear about your astrological chart is also meant to be interpreted intuitively, by what does or does not feel good for you. So before taking any astrologer's word as truth, always ask yourself if it feels right for you!

Different Types of Astrology: Western vs Vedic

As mentioned in the history of astrology there are other ways to interpret the planetary placements in the sky. One way of looking at it is through Vedic Astrology. The reason I want to touch on this is because Vedic Astrology has very similar signs and houses to western astrology, but the interpretation is completely different. I have been using Western Astrology for more than 20 years now, and I have found it to be very accurate, that is why I have created this book on Western Astrological principles. But despite this, many clients ask me about Vedic astrology, so here is some clarity for you...

Vedic Astrology was developed in India and it is still practiced in many parts of India today. It is also called 'Jyotish Shastra' which translates from Sanskrit to mean "the science of light".

When they speak of light, they are referring to the light coming from the sky in the form of stars or planets.

Although Western and Vedic astrology are very similar the key differences to look out for are the understanding of the position of the star constellations. The Vedic astrologists have a different view of where the zodiac signs are in the sky and this is due to different sidereal or tropical interpretations of the earth's relationship to the sky. Vedic astrology also puts a strong emphasis on the moon and its place in the sky whereas Western Astrology is based on the position of the sun. There are two extra additions that Vedic astrology uses. These are Rahu and Ketu and they are said to calculate the karma of human beings, but Western astrology does not include these shadow planets. Looking at the history of these two astrological interpretations, Vedic astrology is connected with Ayurveda, the life science of Ancient India, whereas Western astrology comes from a history of Babylonians and Ancient Greeks.

The biggest question that is asked in relation to these two versions of Astrology is – which one is the most accurate?

There is no right or wrong answer, and there may be times in your life where you switch between them. The astrologist that reads your chart will give you the most accurate reading according to their beliefs and you are the one who chooses whether to take that information on board or not.

I invite you to read through this book, apply the knowledge to your own chart and life, and then make a decision about how accurate you think the Western style of Astrology is. I have a feeling you may be pleasantly surprised with your findings!

How to use this book

This book is designed to be your very own handbook and beginners guide to understanding astrology. A few tips before you begin...

I recommend getting a small notebook that you will dedicate to your astrological discoveries.

As you work through the book, you will understand more about yourself, your life and the way you do things. Write these down and refer back to them as your understanding deepens throughout the book. You should also go online and find a free birth chart to download. If possible, have this with you as you read the book. Whether it is a printed or digital copy, having your chart visible during reading will allow you to apply the practical knowledge in this book directly to your own chart.

As you make your way through the book, you will learn more about the *Four Elements*, and *Three Qualities* of all 12 Zodiac signs.

In Chapter Three, you will see an overview of the 12 Zodiac signs. Most people are already aware of the 12 signs due to the popularisation of horoscopes; however, you will quickly learn that you are more than just your sun sign. This will become more apparent the more you discover about the other elements of your chart and the impacts they have on your personality.

Moving into Chapter 4, the focus shifts to the great 'Gods of the Universe', also known as the planets. You will learn the ancient Greek stories of the personification of the planets and the vastly different qualities that each of these celestial bodies possess.

In Chapter 5, we will discuss the 12 houses. The houses are the sections of your birth chart that hold a unique energy and meaning. At this point, as you align the houses with your birth chart, you will gain a lot more clarity about your astrological journey. My clients often report this consolidation of the houses with their birth charts as being the pivotal point where their entire chart clicks into place, and everything starts to make sense!

Now, by Chapter 6 you may be feeling a little overwhelmed, but that is natural. As with learning any new skill, it takes practice and patience to improve. To make it easier, refer back to your notebook and begin to write down which sign and planet is in each house. There is a step by step guide to reading your chart in Chapter 6, so follow it closely and record your observations.

If you have managed to put all of the information together and are feeling excited to learn more, the final Chapter – finding the perfect lover – is the perfect ending to the astrology journey. In this part of the book you will learn about compatibility and what makes the perfect partner for you. It is fun to read through both your own and your current or potential partners' sun signs and see where there will be a good flow and where there might be challenges. But remember, there is more to one's astrological make up than just the sun sign, so don't go changing relationships after reading the book! Always use your intuition to identify if you are in the right place with the right person.

After you have reached the end, it is time to flip the cover and read again! This book is an invaluable resource that can be referred to as often as necessary to uncover the astrological truths about yourself and others. Have fun with the work in this book, use it where it feels good to explore and remember that even if something is written in the stars, you have the emotional ability to choose how you will react to it – whether you see it as a blessing or a lesson, there are no negative outcomes!

Conclusion

Every time you look at your chart you will discover more about yourself and the world. Even though there are different ways to interpret the messages of the stars, trust that you will always receive the right message that you need to hear. Through a long history of sharing wisdom and stories, the stars are ready to speak to you. This book, written on Western Astrology, is designed to be a companion to beginner astrologists, so that you can start with the simplest yet most effective tools to put this ancient art into practice.

So find your chart, grab your notebook and brew a cup of tea. Get ready to start deciphering the messages of your soul!

---

1. *Nelson, J, 1952. Planetary position effect on short-wave signal quality. Electrical Engineering, 71(5). 421-424. DOI: 10.1109/EE.1952.6437478.*

Chapter 2

Astrology 101

Throughout this chapter you will begin to understand the qualities that each of the 12 signs of the Western Zodiac hold. These 12 signs can be broken down into smaller categories that create a unique and descriptive way to understand each of the signs better. The main ways that we will be categorising the signs is through the elements and the qualities, as well as by grouping them into polarities. The more you begin to understand how these four elements, three qualities and two polarities intersect and intertwine, the more you can begin to look deeper into the ways astrology works. The four elements are Water, Fire, Air and Earth and each holds its very own characteristics. There are three Zodiac signs that are connected to each element, and we will explore this connection and what it means throughout this chapter.

There are also three qualities, known as cardinal, fixed and mutable, which again hold very specific energies that can be used to describe the way the Zodiac signs work. The three qualities relate to the time in the season at which these signs begin.

The polarities are split into positive and negative, and they are commonly compared to introverted and extroverted energy. Understanding the polarities will give greater depth to the elements.

You can often see the different elements and qualities reflected in your friends and family if you know their sun signs or rising signs. As you explore the information in this chapter, see if you can relate it to someone that you know! Just as every single person has their own characteristics and personality traits, so too do the Zodiac signs. Perhaps you even see yourself being reflected in some of the descriptions here. So, let's begin with the basics.

The Four Elements

The four elements are the essential foundations of nature itself. We as humans are a living and breathing form of nature in motion, so it is no wonder that we each hold these elements within us. Water is in every cell of our body and it is always flowing in our blood. The transformative nature of fire is seen in our digestive system as the food we eat is converted to energy. Air is our breath, constantly moving in and out of our bodies. Earth is our physical form, the bones and muscles and structure of our bodies.

These elements are seen in every person but to varying degrees. For example, someone might have a strong digestive system but weak lungs, meaning that they are strong in the element of fire but weaker in the element of air. As we look at each of the elements in depth you will begin to see where your strengths and weaknesses reside. Some of these traits are inherited by your current lifestyle, and healthy improvements may be made to change them. But often you are born with these specific strengths and weaknesses as part of your character, and it is reflected in your astrological birth chart.

One note before we continue - it is important to remember that no weakness or challenge is ever permanent. If you feel that a part of your chart is creating a strong imbalance, there are many remedies to bring yourself back into balance with movements, food choices, herbs and other holistic health practices.

Water

To be connected to the water element means to go with the flow, to swim in the depths and to dive into mysteries, knowledge and wisdom. The three Zodiac signs that are connected with the water element are Cancer, Scorpio and Pisces. These signs are in touch with their emotional intelligence, they are very intuitive, and they also have an air of mystery that surrounds them.

The main characteristics of water signs are their ability to know their emotions, and usually they don't have any problems displaying these emotions either. The downside of this is that they can easily become too attached to their emotions and let it rule them, often feeling out of control of how they think and feel. They may also be very attached in relationships, whether this is friendships or romantic relationships, because emotions come before rational thoughts.

These signs are also highly sensitive, meaning that they are usually aware of the energy around them and can easily become overwhelmed if this energy is too much to handle. They may get easily upset or hurt if negative conflicts arise, whether that is in their own life, or someone else's. With high levels of empathy, they are great at supporting others, but they must also remember to take the time to clear their own energy so that they can maintain good mental health.

Deep and intuitive are words that are often used to describe water signs. With their high level of emotional intelligence and fine-tuned sensitivity it is almost impossible to lie to these signs because they will see right through you. Water signs *feel* a situation rather than think about it, so if their gut feeling is giving them signals, they are going to listen to it. Trying to manipulate a Cancer, Scorpio or Pisces will not end well; their intuitive alarm bells will be ringing loudly!

These personality traits make water signs very home and family orientated. They care deeply about the ones that they love and will go to great lengths to make sure that it is known. They are good listeners, and may help others get in touch with their emotional side too. If you are a dear and trusted friend of one of these water signs, you can know it is a friendship that could last a lifetime.

Water signs in love like to go deep, of course. They fall quickly in love if they feel that the person is intuitively right for them. They are easily attached to their partners, which can be a sign of loyalty, but also a sign of needing too much emotional support. If you are a water sign, it is important to create healthy boundaries in your relationships so that your partner doesn't feel too smothered by your love. Luckily for water signs, they tend to naturally attract loving partners. Their intuition never lies.

When it comes to money, water signs want to have security. They are good at saving money so that they always have a backup. Finances can be an emotional topic for them, but that's because everything in their life has an emotional attachment to it. If money is scarce, or income is too big, then overwhelm and anxiety can arrive. So, if you are a water sign, it is best to consult a trusted accountant to help you, someone who is probably an earth sign with a more practical money mindset!

Health is important to water signs. First and foremost, the body, mind and spirit must be in sync, for this is what keeps them aligned with their intuitive side. As mentioned, water signs always explore the deepest depths of understanding, and have a natural pull toward exploring spiritual connections. This constant search for more information and a greater understanding is what keeps their minds and bodies healthy. When looking for a home, these signs will only settle in a place that suits them. Usually, they feel comfortable and settled near their natural element, so choosing a place near a natural water source is important to consider. Security is also extremely important to water signs. They need a place where they feel safe and settled.

Movement is essential for water signs as they need to keep the flow of their emotions moving through the body rather than becoming stuck and stagnant. Sports such as yoga, swimming, diving, and dancing are great choices for the water signs as it creates a gentle flow through the body whist also allowing movement for their stores of emotional energy.

If you are a water sign, you need to ensure that you are finding healthy ways to balance your emotions. As well as yoga, practices such as breathwork, tapping and meditation are incredibly healthy ways to return to your centre. Sport and movement should also be done regularly to loosen up the body and to enjoy the flow of life.

Water signs have a softness yet sureness that is a trait to be admired. They impart their intuitive wisdom onto others, and everyone is appreciative of their emotional gifts!

Fire

Fire is the element that creates change, it is a driving force and it moves with gusto. As a result, motivation and determination are two of fire signs' greatest qualities. The three Zodiac signs that relate to the element of fire are Aries, Leo and Sagittarius. These three are dynamic, passionate personalities with a strong desire to explore people, places and things.

The major traits that characterise a fire sign are the dynamic and active energy that they embody. Fire is always a little spontaneous, and it burns with intensity, and people who are ruled by a fire sign hold these same qualities. Although both men and women hold the energy of fire, this element is more commonly seen as masculine, for it is the energy of Yang, and is a driving force behind getting things done.

Fire signs' dynamic energy enables them to be talented at many different things. Their impulsive behaviour forces them to seek and learn new skills and new ways of being. These signs are constantly giving in to their restless energy, always moving locations, and having the urge to travel on a whim. If you look into a fire sign's eyes, you will see the passion that lives there; passion for adventure and for new landscapes. Although this can be a wonderful way to live, it also creates a very temperamental attitude, making it difficult to keep up with a fire sign. Sometimes their lust for new places can be exhausting, especially when paired with their intensity and determination to achieve their travel goals.

When a fire sign falls in love, the world will know about it. These signs *love* public attention and affection. They want to be sure you to know how much they love you, which may result in grand gestures or romantic holiday trips together. Fire signs are passionate in love and will usually seek an equally adventurous or mysterious partner to feed their sense of curiosity. They don't know how to move slowly and will often dive into relationships quickly and intensely. Fire signs, if this is you, it is important to try to go slow sometimes and to take careful steps to avoid burnout.

Fire and money burn well together. Fire signs are big spenders but also good at always having enough to spend.

If something catches their eye that they love, there will be no hesitation in buying it. Listening to their gut feeling, fire knows how to spend well and stay savvy with money. The natural talents of fire signs often bring money into their life in a variety of ways. Their creative drive will always keep their wallet full!

Looking at the health aspects of fire signs, one thing that is in no scarcity is energy. As mentioned before, energy levels can be constantly high, but it is important that fire signs learn how to manage this. Burnout is a big problem for fire signs as they often find it difficult to tell when their own energy is running low, they only seem to notice when it has already hit the floor and then it is too late.

Physical movement in healthy ways is also important for fire signs. High intensity activities such as running, team sports and interval training are a great way to use their energy. However, if they want to avoid the intensity dying down, it is important to have a different variety of movement in their lives. Competition is equally important for motivated and determined fire signs. At the end of the race, it will most likely be two fire signs fighting it out for first place. Fire likes to win and can become impatient with themselves or others if this is not happening. To stay balanced, it is best to find a dynamic fitness routine and daily movement.

It is also important for Fire signs to keep on top of the amount of projects they are taking on. Since they tackle all tasks with passion and ferocity, too many projects can cause overstimulation and a quicker route to burnout. This makes it highly important to avoid overscheduling. If this applies to you, make the most of using a calendar, ensuring you are planning with a sense of *practicality* rather than *ambition.*

Fire signs are already naturally stimulated, so external stimulants can very easily push them over the edge. Avoiding excess amounts of caffeine, sugar, and spices in their diets can help restore balance to agitated, overstimulated Fire signs.

Fire signs have big hearts, and their loud and proud attitudes often entertain those around them. However, it is important that they conserve their own energy to make space for their personal drive and desires. With practice and patience, Fire signs can find the perfect balance: excelling in personal endeavours, while inspiring all around them with their generosity and love.

Air

Moving like the wind, spontaneous, unpredictable yet sure, air signs have a mind of their own. Gemini, Libra and Aquarius are the three air signs that enjoy conversations and stimulation of their intellectual minds. If you are looking to inquire deeply into a topic, then chat with an air sign, for they will continue investigating with thoughts until new insights and revelations are revealed.

The mental mind is what rules the air signs. Their cerebral cortex in the brain is often active and this helps them to analyse and investigate ideas and theories. These signs could talk forever, so if you enter into a conversation with one, know that you are in it for the long run. Their focus is on communication, and they like to share their ideas with others. They embody a yang energy, making them active in their pursuits of knowledge. For them, knowledge is power, and they can usually talk themselves out of any tricky conversations. With the quality of air being light, their minds are also up in the air often and they spend a lot of time with their own thoughts, drifting away with the clouds. This creates a sense of them not being present and it often takes another sign – such as fire or earth – to bring them back down to the planet. With this lack of grounding, the air signs come up with great ideas but unfortunately not many of them come to fruition until they pair up with another person who is willing to help them get the work done.

In love, they are also often up in the clouds, and they require stimulating conversations and mental stimulation from their partners. Without someone who can match their level of intellect and mental curiosity, they will become bored quickly. But aside from deep conversations, air signs will choose a partner that shares the same sense of humour. If you are in a relationship with an air sign you can expect a lightness and free-spirited approach to life. They will never rush into a relationship, for they need time to analyse their partners before they fully commit. Even once committed, you may find their free-spirited approach wants to stay open to new options and ways of being in a relationship so it's important to be flexible with an air sign in love.

Air signs are also able to make good choices when it comes to money. Their good ways with technology often help them in this topic, for they can easily turn to mobile apps to help them stay on track with their finances. They love to spend most of their money on social events, for they are very social people, and they also don't mind a little impulsive spending sometimes. When something feels important to them, air signs won't mind spending money on it.

In regard to health, the air signs usually focus on their mental state rather than their physical body. This means that they need to look after their mental health through expressive communication. Personal relationships are helpful in this sense, for the air signs need close friends and family that they can trust in to feel safe when expressing their thoughts and concerns. When they are given the opportunity to express themselves, their mental health will thrive.

Movement is also great for air signs as it grounds them back down into their physical bodies. Guided workouts, yoga classes, and gym classes are great for keeping them active and working out with a friend is great for accountability. If they have a workout partner, they are more likely to actually do the exercise rather than just think about it!

When an air sign is out of balance, they can feel very ungrounded, agitated, and anxious. This makes grounding exercises the perfect solutions for these feelings. Earthing, which is the act of connecting your bare feet or parts of your body to the earth, is great for an instant grounding experience. Spending time in untouched nature is also a powerful way to reconnect with the earth. Breathwork is another great practice to incorporate daily for air signs, for it helps to move stuck energy in the body and to clear out old thoughts and feelings.

Air signs will always seek meaningful connections and good conversations as this keeps them balanced and gives them nourishment for their mind.

Air signs are wonderful friends, for they will show loyalty through deep and meaningful conversation and connections. They will take time out of their day to chat with you whenever you need help. You may just need to call them first to bring them back down to earth!

Earth

Strong, steady, and grounded, earth signs will make you feel safe in their presence. Taurus, Virgo and Capricorn are the physical beings of the earth element. They care deeply about their bodies and their physical health, and they take their time when it comes to all kinds of occasions.

You will know an earth sign from their practical approach to life. There is no spontaneity here, rather routines and pre-made plans will fill up an earth sign's day. They are in touch with their physical body, and they enjoy their sensual nature. They like to wear clothes that feel comfortable and nice on the skin, and they like to eat to nourish themselves. You will never be able to make an earth sign rush anywhere because their grounded nature only knows how to move slowly and steadily. They do things thoroughly and they like to consider all of the little details as this helps them to create a solid foundation for any ideas or projects that they choose to pursue.

Earth signs are classical yin energy, meaning that they only move when the time is right. They are the kind of sign that wants to make a safe home and will not move from that home for the rest of their life. They hold a lot of feminine energy which makes them a calm, soft and loving personality. The feminine and yin side in them, makes them out to be a great 'mothering' figure as they care deeply about the physical and emotional needs of their loved ones.

They are the most steady and stable of all the signs making them a dependable person that you can trust fully. They embody this quality in love as well, meaning that they are a dependable lover. If you have an earth sign as a partner, it is unlikely that they will leave you or make any big changes without you knowing about it. Their attraction to others is driven by their senses. Their sensual nature means that they feel into a person or situation, taking great care and caution to decide if it is the right relationship for them. Physical cues are important to them, so taking time for touch and physical support will help you to convince an earth sign that you love them. Of utmost importance to them is that their partner shares the same values and is also driven by physical attraction and connection.

In regard to the health of earth signs, they need a set routine to stay healthy. This includes a healthy sleep schedule, and it is crucial for them to stick to it. If they lose track of their routine or need to change it for a few days, they will really feel the impact on a physical level. As they like to see results, it can be beneficial for these signs to do sports or activities where progress can be tracked. This physical progress will keep them inspired which makes yoga, running or weightlifting great choices for these signs. It is important for them to find one or two types of workout that they love and sticking with them by incorporating them into a weekly plan or routine. This will encourage the earth sign to actually follow through with their workout goals.

To bring balance into the life and body of an earth sign, routine and structure are key. These signs have a tendency to get stuck in bad patterns or behaviours and it can be really hard for these signs to break a habit. That is why a detailed plan is important for earth signs so that they have the physical motivation to follow through. Engaging in different activities is also a great way for them to avoid falling into laziness or stagnation.

Earth signs are reliable and able to bring you back to your own self and physical body when you need it. A loving hug from an earth sign is like a spoonful of medicine, for they give with full hearts and kind intentions when they choose to be your friend.

The Three Qualities

Aside from the elements, the Zodiac signs can also be grouped into three qualities, these being cardinal, fixed and mutable. Every sign is influenced by one of these qualities and it is determined by where they fall amongst the seasons. The beginning of the season is the cardinal quality. For example, the start of Libra corresponds with the very beginning of Autumn (in the Northern Hemisphere) or Spring (in the Southern Hemisphere). Mixed signs are those which fall right in the middle of a season, like Scorpio. Finally, as we come towards the end of a season, the signs take on a mutable quality. For instance, Sagittarius marks the end of Autumn/Spring, tying up all the loose ends and lessons that occurred that season! These three qualities represent how the Zodiac signs relate to the world around them. Understanding the three qualities adds an extra layer to the elemental descriptions above, providing in-depth knowledge about the way the Zodiac signs influence your life.

Cardinal Signs

Cardinal signs are those that start at the beginning of a new season. They bring with them a fresh start, and they face the world with new and inspiring ideas. The signs that are Cardinal are Aries, Cancer, Libra and Capricorn. They are innovative and impulsive, making them unafraid to pursue new ideas or to go where no one has been before. They are big idea generators, and often good to chat with if you are feeling stuck in a situation. The way they shed light on a situation is unique and can be of great benefit to others. Once they have their ideas, they will push strongly to get it started. Within no time, their ideas will be up and running and this will continue to lead to new innovations.

Fixed Signs

Taurus, Leo, Scorpio and Aquarius are the four fixed signs. This means that they are the main body of the seasons, and they carry the weight of whatever that season brings. This makes them capable of endurance, stable in their emotions and focused on their goals. If you need help with planning an idea, or staying on track to reach a goal, consult a fixed sign and get their tips and tricks of how to maintain integrity for the long run. The downside to fixed signs is that they are often stuck in their ways and routines, and they need to be shaken up a little to have the urge to try new things.

Mutable Signs

Mutable signs fall at the end of each season, making them prone to constant movement and change. Gemini, Virgo, Sagittarius, and Pisces are mutable, and this makes them highly adaptable to new situations. They are flexible in their thoughts and their bodies, and they are often creating space for new things to come their way. Unlike the cardinal signs, who think of the ideas themselves, the mutable signs just let the ideas come to them. They trust deeply in the ever changing world around them and have no problem staying open and adaptable to the flow of life.

Polarities

Another way to understand the Zodiac signs is by breaking them up into polarities. There are two ends of polarity, those being positive and negative. The four elements can be found under one of these categories. Fire and air are at the positive end of the polarity scale whereas water and earth sit at the negative end. Let's have a look at what this means for the signs.

Positive Signs

Positive polarity indicates that energy is expressed outwardly. This looks like a bold, confident and extroverted person. People with fire or air signs tend to fall under this positive polarity. When you think about fire, it is bright and unafraid to illuminate the room or environment it is in. Fire can be larger than life, raging and dancing for attention. The element of air is also similar. Air is felt when it moves into a room, it is noticed by the way it makes the trees sway or the way a cold breeze brings goose bumps to the skin. It is also unafraid to be seen or heard. These two outgoing elements are connected to the positive extroverted type of polarity.

Negative Signs

Negative on the other hand, suits the elements of water and earth. Water is cool, calm and collected. Yes, there can be raging storms and wild seas, but more often than not, water just flows downstream, doing its own thing. This is the introverted type of energy that corresponds with negative polarity. Earth is also an introverted energy, taking its time to make decisions and coming to its own terms in its own time. Earth signs are grounded and stable in the way they think and move, giving them the time to have their own introspective conversation with themselves before expressing anything outwardly. These introverted elements are the ones who turn inward to gain their energy rather than to seek it externally.

Conclusion

There are many ways to classify the Zodiac signs, and this is why it is important to give each of them unique elements, qualities and polarities. You will never just be one element or one quality, but a combination of many of these. This is what makes every person unique in the way they live their life and how they express their personality. The more you can begin to understand these elements, and qualities that are present in your birth chart, the easier it can become to understand the essence of your true nature. This will help you to apply some practical tips in your life to use your elements and qualities in the most beneficial way. There is always more to learn when it comes to understanding astrology but being able to use the elements and qualities are a great foundation. From here, your knowledge can build and expand as you begin to explore the rest of your chart!

Chapter 3

Star Signs

The 12 star signs are the most basic part of astrology that most people know about or have at least heard of. Sometimes the "star signs" are also referred to as "sun signs" as they actually correspond to where the sun is in relation to the stars at any given moment. Star signs are the most commonly known part of the astrological birth chart for it is the easiest aspect to find out. All you have to do is look at which sign the sun was in at the exact time of your birth. This is usually easy to work out as the sun moves slowly in the sky, staying in one sign for around 29 days. So, for example, if you were born when the Sun was in the sign of Taurus (April 20 – May 20) then your star sign (also sun sign) is Taurus! Simply knowing your star sign can already tell you a lot about yourself. But to understand why, we must look to the qualities of the Sun. The sun is a sustainer of life, one that illuminates the path ahead and drives out darkness. In this way, the sun is the sustainer of our inner world. The Sun illuminates the truth of our conscious mind which highlights the way that we want to live and how we will display our creative life force. The Sun also gives us a hint as to what style of living our inner self craves. For example, an Aries would like to live on the move, forever exploring and changing locations, whereas a Taurus may want a steady and stable base with comfort and familiarity. When you look deeper into your Sun sign, you will be given these hints as to the lifestyle that best fits your inner nature.

Many astrologists relate the sun to the 'ego', meaning that the sun is also one of our biggest drivers. It influences the way we think and how we act in front of others. When we align our actions with the energy of our sun sign, we tend to feel more balanced and happy in life. So let your true personality shine and note which of the qualities of your sun sign you are already embodying.

Star signs are easy to work out, because all you need is the birth date, which means you can also read the information about your friends and loved ones. See how these descriptions fit into your life and the lives of others! Have some fun with this as you remember your own true nature, that of your inner Sun!

Characteristics of the 12 Zodiac Signs

- **Aries (March 21 – April 19)** Aries is the first sign of the zodiac, making it a sign that naturally wants to come first in everything. Full of fire and represented by the Ram, Aries personalities want to charge forward through life. Their competitive nature makes them charge with their head first which makes them appear as explorers, unafraid to go into unknown lands. Albeit they often make these moves without stopping to consult their friends or family, and are sometimes seen as a little self-centered. This is due to the fire element that lives within them, which makes it easy to burn away the opinions or judgements of others and follow their own inner calling toward the next right thing. Aries are dynamic and eager to pursue their next passion and you will often see them intensely focused on the task at hand. They are quick to master a skill, but just as quick to move onto the next one. If you can move past their explosive temper and impatience, you will get to experience the playfulness that an Aries brings to those around them. They are mostly positive and always living life to the fullest, making them a great companion to take adventures with!

- **Taurus (April 20 – May 20)** After the fiery heat of Aries, Taurus comes into play. The bull also charges ahead in life, but not without contemplating which path is the best to take. Slow to move and stubborn when it comes to change, Taurus is an earthly creature. Fond of earthly pleasures and comfort, Taurus likes to surround themselves with luxury. A steady homebase and a stable life is one that is ideal for a Taurus. You can forget about quick decisions, but you can rely on long term plans. Taurus likes to think ahead and will be well prepared for any upcoming events. When in balance, they are in touch with their bodies and their creative side brings forth a strong desire for them to be themselves. They have a sensual flare to their personalities, enjoying physical delights and also not being afraid to go out and get them. Taureans know that luxury comes at a cost, but these Bulls are willing to put in the hard work for the long term so that they can receive the big rewards! Once they have made up their minds about something, there's not much point in arguing. On the soft side, Taurus is a dependable and loyal friend with love to give.
- **Gemini (May 21 – June 20)** Gemini is represented by the twins, which has been chosen

for a reason. There is more than one side to a Gemini, and you have to really know them well to understand this. Smart and passionate, those with Gemini as their star sign are great conversationalists. In fact, it's often hard to get them to stop talking! Energetic and quick-witted, Gemini will be able to blend in with any person or event. Their multifaceted personalities earn them the reputation of being the "chameleon" of the zodiac for they are so great at being up to date with whatever conversation is happening. With so much mental energy, they are constantly thinking up new ideas and searching for more time to get it all done! Geminis are curious and progressive, and their love for communication keeps them busy chatting, texting and posting on social media. But you have to be quick to keep up with a Gemini, for they are always moving ahead with their plan to transform the world. With their playful and intellectual minds, Gemini will plant new seeds in your mind and will tell you the steps you need to take to make them grow!

- **Cancer (June 21 – July 22)** Crabs are both masters of the oceans and of the sandy shores, making Cancer a sign that can easily navigate between both emotional, watery depths and grounded, earthly pursuits. When Cancer chooses to go deep, there is a natural sense of intuitive knowledge that lives within them. They are connected to the energetic and psychic realms of emotions, yet practical when it comes to physical safety. They are highly sensitive to both their surroundings and other people and this is what sometimes makes them withdraw. With its home on its back, the crab always has a safe place to return to and this is one of the most important needs for Cancer, a safe place to retreat when they are feeling overwhelmed. This sensitivity can make them come across as distant or cold, but you must know that they truly are caring, warm and gentle at heart. That being said, it can take a while to get to know a Cancer, but when you are welcomed into their warm and cosy home, you know you are a trusted ally with whom they will share their compassion and mystical side. Cancers are great at attracting the right friends and lovers into their life, so if you are close to a Cancer, you can trust that you are in the right place.

- **Leo (July 23 – August 22)** Proud and big-hearted just like a lion, Leo is a sign that is unafraid to be in the spotlight. They strive to have the loudest roar and the most beautiful mane, so that everyone around them knows that they are in the presence of a Lion. The personality of a Leo is dramatic, and they like to be the center of attention, captivating others with their charisma and warm sense of humor. The fire in their hearts makes them generous givers and if you have Leo as a friend, you can be sure that they will be loyal for life. Their dedication to their own goals and those of their friends, makes them a stable place to seek advice from. They stay strong until their ego gets in the way, which has the tendency to bring pride and jealousy to the forefront. Leo's ambition and drive will take them to new places and show their true strength. If you need to be brave, consult a Leo, for they are the masters of bravery, with undying amounts of physical, emotional, and mental strength. Take a dose of their optimism, and you too will be thinking you can take on the world!

- **Virgo (August 23 – September 22)** Virgo, the virgin, is a practical sign thanks to its earthly influences. They have a logical outlook on life and it makes them a valuable resource to turn to when you need a plan. Step by step, Virgo makes things happen. With the year already fully booked and plans galore, there is not a lot of room for spontaneity in the life of a Virgo, but there is time for loyalty. With patience and perfectionism, Virgos will perfect the skills that they need in life, for they can have unwavering commitment when need be. A great communicator and with emotional stability, Virgo is able to bring complex concepts into simple and understandable terms. They also seek simple beauty, and are still learning that imperfections can be beautiful too. Supportive by nature, Virgos are natural caregivers. They use their kind, gentle and loving ways to solve problems and their resourcefulness to create comfort in the material world. This earthy sign is hardworking and committed to results, making a great teacher or healer archetype.

- **Libra (September 23 – October 22)** Libra is a sign concerned with balance and returning to equilibrium. They are very 'heart-centered' people, wanting to find harmony in all areas of life, including love. With their big hearts, they are able to negotiate good deals that are of benefit to both the buyer and the seller. They are ruled by air, and so they are usually living in their minds, thinking about all the potential options that exist. Once they get fixed on one idea, it can be difficult to change their minds. Social creatures by nature, Libras like to talk about their problems and solutions out loud and seek other's opinions so that they can come to a more rounded conclusion. They have an artistic flare to them, which makes them attracted to inspiring and creative surroundings. Ruled by love, and in the search for companionship, Libras will share their friendship with those that they trust.

- **Scorpio (October 23 – November 21)** There is often a little fear and skepticism in the air when it comes to Scorpios. And for good reason. This sign is one of mystery, and beneath their steady gaze, it is difficult to tell what schemes they are plotting. Unafraid to talk about the deep things in life, and ready to sting with their tongue if they disagree, the Scorpio personality often needs a little warming up to become comfortable around. Their lack of fear makes them brave and resourceful, willing to chase their deepest desires. Their passion is contagious and they value the truth, so if you need to uncover a mystery, talk to your Scorpio friend. Beneath their tough façade, there is incredible depth of emotion, for they do swim in the psychic and emotional realms. It is often their need for power that puts them a few steps ahead of you in every situation. If you can handle the intensity and unpredictability of a Scorpio, know that they will have your back in any situation!

- **Sagittarius (November 22 – December 21)** Back in the realm of fire, Sagittarius is a spontaneous sort. It's hard to predict what their next move will be, as it could be anything from moving houses, to changing jobs, or pursuing a new hobby. Often, they don't even know themselves until it is already happening. They are an extroverted sign, that is fun and loving to others. They love their friends fiercely and are always optimistic, knowing that everything is going to work out in the end. A Sagittarius has many pursuits at once, chasing adventures, new horizons and new insights about life. It is their intense curiosity that drives their passion for knowledge and creates a kind of intensity that can be inspiring to be around. They aren't afraid to enter new terrain or to ask the questions that others wouldn't dare to ask. These fiery signs attract abundance when needed, and they use their resources to travel far and wide and pursue new interests.
- **Capricorn (December 22 – January 19)** If you are going to chat to a Capricorn, you need to have time. Time is the most valuable resource for

a Capricorn, and if you share time with them, then they know that you truly do care for them. Their analytical minds and thoughtful processes create plans that will last. They set their goals and plan every step along the way, knowing that patience and perseverance will pay off in the end. One of the most resilient of the signs, Capricorn will push forward through challenges and pain, overcoming any obstacles that are placed in their path. They have practical minds and a plan at hand, with a long-term goal that they are striving for. Their unwavering determination pushes them through and focus is something that comes naturally for them. The perspective of a Capricorn is often created around what it already knows and the lessons it has learned. It carries this knowledge with it through its lifetime and uses it wisely to overcome future challenges.

- **Aquarius (January 20 – February 18)** This sign is the ultimate free thinker, the one who is always on the verge of leading a revolution. Aquarius is the dreamer who envisions a world where all people and creatures are free to do whatever they like. An open-minded character, Aquarius has humanitarian needs at heart and will do whatever it takes to make this world a better place. If you need an original idea,

Aquarius is sure to have one. They are imaginative with their mind and deep in their thinking, often creating ideas that no one has had before. As an air sign, they are very social creatures, sharing their thoughts and ideas with those around them. This is in the hope that they will connect with other like-minded revolutionists! With bountiful inspiration, creativity and innovation, this sign is progressive in its ways and passionate about change! They do have a sensitive and soft side too, but only when they return from their dream state do they feel this humanness setting in. This is why it is not surprising that they like to live in the mind and spend so much time in the dream states.

- **Pisces (February 19 – March 20)** The last sign of the Zodiac is the wise and intuitive Pisces. With the ability to swim in oceans of emotions, Pisces sifts through the psychic realms with ease, swimming with delight in the energy of others. This makes them highly empathetic characters and able to relate to others easily. A Pisces will often sense how a person is feeling before they even know it themselves. This is because Pisces is the last of the Zodiac signs, meaning that they have already journeyed through the Zodiac and absorbed all of the lessons they needed to learn.

They are now left with the wisdom and knowledge of the Zodiac and this makes them affectionate to others as they make their way through the lessons of life. Pisces hold great amounts of compassion in their hearts, and they trust in the magic and mystery of life. Often captivated by illusions, this sign needs to remember to return to the land sometimes, and use their psychic knowledge for practical pursuits

LOGOS OF THE 12 ZODIAC SIGNS

Aries The symbol of Aries is the Ram, a strong, bold and courageous creature. Aries is said to be a golden ram that rescued Phrixus, a mortal prince by taking him away to the Gods. Then, Phrixus sacrificed the golden Ram for the gods. Later, when Phrixus was once again facing death, a golden ram with wings came by to save him. This shows that Aries is unafraid and takes initiative in challenging times. It is a story about the ram as a steadfast and brave character, true to the courageous nature of Aries.

Taurus

Taurus is the Latin word for Bull and it has been identified in the skies since ancient civilizations. To the earliest star gazers, the Bull represented strength and war but also steady and reliable love. Perhaps you have heard the word minotaur – which is a Greek-Roman mythological creature that is half human and half bull. This is how the humans integrated the strength of the bull into their own lives.

In Ancient Greek mythology, the God Zeus, fell in love with Europa, and turned himself into a magnificent white bull to win her love. When the princess got on the back of the Bull, she was kidnapped by the undercover Zeus, and then Zeus put the image of the bull in the stars to be forever remembered. Here the bull remains in the sky to this day.

Gemini

The Gemini symbol consists of the roman numeral two, with additional branches above and below. This symbol is said to represent companionship, as it is two lines side by side, joined as one. In Roman mythology, the two parts of Gemini are for the two stars Castor and Pollux, who are also twins. These two stars have an unbreakable bond. As the four lines of the Gemini symbol all go in opposite directions, it is said that this is why the energy of Gemini is so scattered. But when tuning into the center, the essence of Gemini is held safely within the shape.

Cancer

Cancer is the Latin word for Crab and this shape has been observed in the stars since times of Greek mythology. The crab in the sky was said to have pinched Heracles while he was fighting. As revenge, Heracles had crushed the crab, and then fought the enemy. At the end of battle, the crab was placed in the heavens in the form of a constellation.

Leo

Leo the Lion has been spotted in the sky for many thousands of years. The civilizations of Ancient Mesopotamians, Persians, and Indians have all spoken of this Lion in the sky at various times in their culture's history. However, commonly known today, the Greek mythology states that Heracles slayed the Lion that had been terrorizing the city of Nemea. Although this lion had the toughest of skin, Heracles still found a way to end the life of the lion, and in the end that was with his bare hands. This tough skin is a trait of Leo, brave and courageous, it takes a lot to knock Leo down.

Virgo

Virgo is the symbol of the Maiden, for it literally translates to Virgin. The symbol is the letter M, thought to represent Maiden, with a loop at the end. The loop has been interpreted in many different ways, but most include the notion of chastity, relating to the virgin. Some interpretations say that the loop represents the natural cycles of life and death, believing in the karma of actions. Another creative way to look at it is as thought the loop is a woman with her legs crossed, the virgin herself.

Libra

Represented by the symbol of the scales, the symbol of Libra is one of equilibrium and justice. It is said that the Greek Goddess Themis, who oversees divine law, is associated with Libra. Themis uses her scales to remain balanced and to make pragmatic decisions. When Themis' daughter ascended to heaven, she took the scales with her, creating the constellation of balance in the sky.

Scorpio

Scorpio is recognized by the constellation Scorpius, a large group of stars known by the long tail. It is said that the scorpion originated when Orion's brother Apollo asked Mother Earth to kill Orion, as Apollo was a jealous man. After death, Zeus placed the Scorpion up in the sky as a constellation, reminding people of the dangers of jealousy. The symbol that is used for Scorpio is depicted as the letter 'M' with an arrow on the tail end. This arrow is said to represent the ability for a Scorpio to wait until the right time to make their move for either love or revenge. Like an arrow hitting the heart, the Scorpio knows how to hit the sweet spot.

Sagittarius

Half human and half horse, Sagittarius is a Centaur, one who lives in the realm between heaven and earth. With an arrow in hand, known as the archer, the symbol for Sagittarius in the zodiac is a bow and arrow. It is said that the Archer can shoot far and with good precision, making the Sagittarius one to achieve their goals, or hit the bull's eye in every situation.

Capricorn

There is an unusual combination for a Capricorn, and that is the sea-goat. With the head and upper body of a goat and the lower body and tail of a fish, the Capricorn is based on the Sumerian god of Wisdom and waters, Enki. The earth goat features, and the water fish features represent the ability of the Capricorn to move between the logical world and the imaginative one.

Aquarius

Aquarius has an interesting symbol, for it is an air sign, but it is represented by the Water Bearer. This is said to be due to the God Ea, who carries around a vase of water. When he dipped his vase into the Nile, it was said that the Nile flooded which brought abundance and growth to the surrounding towns. The waters of Aquarius are also healing and cleansing and this is seen in the intuitive and innovative ways of Aquarians.

Pisces

Pisces is represented by two fish swimming in opposite directions, yet they are linked together by a middle band. This middle part can symbolize the indecisiveness of Pisces, always tending to live in-between choices or decisions. The band in the middle also represents the bond between Venus and Cupid in Roman Mythology or Aphrodite and Eros in Greek Mythology.

Why doesn't my sun sign sound anything like me?

The sun moves through each of the 12 zodiac signs for a month at a time. This is why there are so many variations of personalities in one sign. For example, you may know a few people with Aries as their sun sign, and even though they may share some similar qualities, of course they are individuals with their own likes and dislikes, passions and hobbies. This is because there is much more to astrology than just sun signs. At the time of your birth, each planet falls into one of the 12 zodiac signs and has an influence in your life. Your sun sign is simply the sign in which the Sun was in at the exact time of birth.

If you look at your birth chart, you will see that you have influences from every sign and every planet. The entire chart is like your blueprint, and it is different for every single person. When you look at your chart, you will see that some signs are more emphasized than others, giving you more attributes from these signs. For example, you might be a Pisces sun sign, but if you have the planet Mars placed in Aries, then chances are you may feel more like a warrior than a sensitive artist.

I WAS BORN ON A 'CUSP', WHAT DOES THIS MEAN?

The cusp is the transitory time between zodiac signs. To be born on the cusp means that you were born near the beginning or end of a sign. Every year, the exact day and time of the signs vary slightly, meaning that if you were born on the cusp, it is likely that your Sun sign may vary to what you had originally thought. The idea that the signs blend together at the cusp was a concept created in pop astrology. The idea was popular because it allowed people to read their horoscope regardless if they knew their sun sign or not. This was commonly used before the times of exact calculations. But today, even a simple search on the internet will allow you to find sites where you can enter your birth date and time and see your very own chart. Then, you won't need to worry about the concept of the signs blending together.

Once you have more clarity on which sun sign belongs to you, you are already on track to understanding the rest of your chart. The sun is often the easiest to discover and read about, but there is a lot more depth and detail available when you have an accurate chart available.

Conclusion

Now that you have your exact sun sign and you understand what that means for you, you are already on the path of diving deep into the world of astrology. This is like the stepping stone into deeper insights about yourself, the world and the stars in the sky!

As you continue your journey through this book, the layers of meaning will start to add up. So stay curious as you move onto the meanings of the planets.

Chapter 4

The Gods of Our Universe

These last few chapters have been about understanding the Zodiac signs, which are an integral part of learning astrology. Often people stop here, happy to know their sun sign, but the depth of your birth chart goes much deeper, and for this we need to look at the planets. When we talk about planets, we are really talking about celestial bodies. Therefore, the sun, the moon and some asteroids are included in the definition of 'planets' for astrological purposes.

Each planet holds its own unique energy, and it symbolizes something in our lives. Usually the planets are described as the mental attitudes we have and the ways of thinking, being or doing that follow. We will take a look at exactly what this means throughout this chapter. This will help you to apply the placement of planets in your birth chart to significant events in your life or personality traits that you have.

The entire study of astrology focuses on looking at the movements of planets and then interpreting these celestial movements to relate to our lives here on earth. Astrology is about understanding the bigger picture of the universe. It teaches us that we are simply just one part of a greater whole, and when we put it all together, the stories and events that are unfolding can be explained by the planets.

How are Planets used in astrology?

Each planet that you read about in this chapter is symbolic for a lesson that we will learn in our lifetime. They rule our desires, our hidden drivers, and our personalities. They influence the way we communicate at certain times, how we show our emotions (or don't show our emotions) to others, and how we face the world when we wake each day. Your own personal combination of where the planets sit in your chart are influenced by the Houses and zodiac signs, and this actually creates your unique personality and lifestyle. So, as you look at your natal chart, you will see a map of the heavens at exactly the moment of your birth into the world.

Greek Gods and the Planets

The Ancient Greeks held great reverence for their Gods, believing that each one played an important part of life here on Earth. The God's were respected and worshipped, and what better way to pledge your allegiance with the Gods than to name a planet after them. This is how the planets got their original names. The Sun:

Apollo, the golden boy of the Gods, the son of Zeus and Leto, was known as the Sun God. As though he was a ray of sunshine himself, he shined his light over the people. He was a protector of children and a picture of good health. It was Apollo that the Ancient Greeks prayed to when asking for better education and health for their children.

Mercury:

As a strong communicator, Hermes was known as the God of Mercury. He was born walking and talking fluently, being able to clearly communicate his needs. He was said to be quick both physically and mentally and he expressed himself through beautiful music on his Lyre, a stringed instrument.

Venus:

When it comes to Venus, people think of beauty, love, procreation, pleasure, and undying passion. This makes

Aphrodite a wonderful ruler of this planet. She was born from the ocean and quickly settled for a top spot amongst the other Gods and Goddesses. She embodies the feminine archetype of love and beauty and spreads this to others wherever Venus lands in their Natal chart.

Earth:

One of the oldest Greek Gods, Gaia, is the Goddess of the Earth. She is said to be the grandmother of Zeus and the mother of the sky. She formed early on in time and has survived and adapted to suit many cultural stories, from Roman times through to modern spirituality. Being strong and steady, she is the ground we stand upon, and the basis of where our journey begins.

The Moon:

Artemis, the twin sister of Apollo (the Sun God) is the Goddess of the Moon. Apollo was golden, and Artemis was silver. She was a hunter with a heart full of wilderness. Sensitive yet strong, she embodies the perfect qualities of the moon. Knowing how to rule her own emotions and yet be sensitive and soft.

Mars:

This fiery planet is ruled by Ares, the Greek God of war. He was said to rule over the more violent and brutal aspects of war and loved to engage in mindless fighting.

This is what gives him the reins over the 'red planet' Mars.

Jupiter:

Zeus, the god of Olympia, of sky and thunder, was one of the chief figures in Greek Mythology. He ruled over Jupiter as well as the 12 gods of Mount Olympus. In charge of everything, he set the tone for right and wrong in the Kingdom of the Gods.

Saturn:

Cronus was the Greek God associated with Saturn. He was the God responsible for 'time' and even today we see his name in words such as chronological or synchronistic! He had the ability to change the seasons and influence the time of day, making him a powerful god and ruler of time based lessons.

Uranus:

Uranus (sometimes spelt Ouranus) was the God of the planet with the same name. He was the ruler of the sky and one of the original gods with an abundance of powers to influence how the future plays out.

Neptune:

The god of the seas, Poseidon, was the god of Neptune. He was a brother to the ruler Zeus, and brother to Hades.

Poseidon had the power to influence storms and earthquakes, as well as the waters and horses. He understood the goals and desires of the Gods.

Pluto:

Far out on the planetary realms, Hades, the god of the underworld was the god of Pluto. He was the last stop that the human mortals would visit on their journey beneath the world. He rules the subconscious and secrets that emerge after a visit to the underworld.

Planetary Rulers of the Zodiac Signs

Every Zodiac sign is ruled by its own planet. This enhances their characteristics and personality traits of the signs themselves. The planets hold a unique energy, each with their own influence on events and aspects of life. Depending on where the planets land in your birth chart, they will influence a zodiac sign more specifically. Knowing the planetary ruler of your own Sun sign is important, as it can tell you a lot about yourself. It can give you more clarity on the lessons you have to learn, the energy you will embody in your lifetime and the mindset that you think with.

Some planets are quick moving and have short and sharp influences, such as mercury, while others take generations to move in the sky, having a stronger impact on society rather than individually. If you have your birth chart, take a look at where each of the planets is placed for a deeper understanding of how they are currently affecting you and your life.

Now, let's look at each planet individually and what they stand for...

The Sun

The sun is the centre of our solar system. It is the planet around which everything else revolves. This also ties in with astrology in the way that the sun is symbolic of who we are at our core - our true self. It represents the centre of our identity and is connected to our ego. It also represents who or what we strive to be in this lifetime and the way that we give our own light to those around us.

In relation to the signs, it is no wonder that the sun rules Leo, for Leo also likes to be the centre of attention and likes for everything else to revolve around them.

Leo is the bright and vibrant lion, unafraid to be the center of attention. The sun also rules the heart and the solar plexus, creating big lovers and good digesters – not only of food, but also of emotions and experiences. When the sun is in balance, it brings warmth, light, and creativity to a situation. When it is out of balance, it may look like a lack of energy or motivation. Wherever the sun lands in your chart, shows you the aspect of yourself that you show to the world. It highlights your shiny personality and how others see you.

The Moon

The moon is the ruler of emotions. It is associated with the 'mother' archetype and femininity due to its tendency to express its emotional needs clearly and publicly. The sun is the actor, and the moon is the reactor, making it one to step out of the limelight and into the depths and darkness. Unafraid to look at the nature of its own being, the moon discovers the best way for you to protect yourself and others. It is no wonder that the moon rules the sign of Cancer, for cancer is the introverted crab that is in touch with their emotions. Wherever the moon is on your chart will show how you express and process your emotions. It also gives you the insight as to how and where you can make yourself feel most comfortable, most at home!

The moon rotates around the Earth, which rotates around the sun and this creates lunar cycles that consists of a new moon, waxing moon, full moon and waning moon period. These cycles hold their own energy and they also have

Mercury

Mercury is the messenger and is the planet of communication. It decides how we choose our words and how we communicate with others. This may be in a friendly, warm way, if mercury is in a positive sign, or a cold and reserved way if mercury falls in a negative sign in your chart. Mercury rules both Gemini and Virgo, giving these two signs the benefit of clear communication. Gemini will never run out of words, and Virgo will never say anything that wasn't meant to be said. They are both signs that are great at coordinating themselves and others, planning and plotting ideas, analyzing tactics and conveying their thoughts.

Look to where Mercury lands in your chart to understand how you process and express information to those around you.

Venus

Venus being the planet of love, rules over the signs Libra and Taurus. This gives these signs an advantage in love due to their easy going and open hearted behaviors. Through understanding where Venus lands in your chart, you will see how you have developed your tastes, pleasures and joys in life. Venus also explains to us how we are in relationship with others. For example, if Venus is in Libra, then you are most likely seeking a relationship with balance, but if Venus is in Aries on your chart, then you are probably trying to seek thrills in your relationships. Venus is about what we seek in ourselves.

Not only love, but also money is concerned with Venus. As it is the planet of comfort and earthly pleasures, it will show how we choose to spend our money in a way that makes us feel good.

Although Venus is commonly seen as a feminine planet, both men and women can embody the Venusian archetype of romantic love and generosity in giving. Notice where you create peace and harmony in your life and see if that relates to your chart!

Mars

The planet of war, aggression, passion and survival instincts rules over Aries. This creates a strong personality in this sign. Aries tend to follow their animalistic instincts which lead them to raw and authentic connections with others. Mars is hot and fiery, which is why it rules in a swift and action-oriented way. Wherever Mars is in your chart, you will probably notice yourself getting in heated discussions around the themes that it is highlighting for you. Mars also relates to one's sexual drives and desires. If you look at the qualities of where Mars is in your chart, you can understand what you seek in a lover (as opposed to Venus who shows us who we seek to be in relationships). It will also show you where in your life you need to be assertive and adventurous so that you can live life to the fullest.

Jupiter

Jupiter is a big and expansive planet, one that instills inspiration and ideas in areas of your birth chart. Ruling over Sagittarius, we can see why those with Sagittarius in their charts are big explorers. This planet seeks insight, wisdom, hope and honor. It reaches out to find purpose and possibility in new ventures. Jupiter is considered to be a lucky planet, one that brings abundance and optimism. Look to where it sits in your chart to see where you will get lucky this lifetime. Jupiter takes around 12 years to travel through all of the zodiac signs, so your luck will change in certain stages of your life, but if you read your chart right, you should be able to see it coming.

Saturn

Saturn is the planet related to restriction, being quite the opposite to Jupiter. Where Jupiter seeks expansion, Saturn finds limitations. But this is sometimes a good thing, for it brings stability and structure into our world. Without boundaries, we would all be dreamers, so it is important to consider where Saturn brings this sense of self-control into your life. Wherever Saturn lands in your chart will be the lesson you are learning for the first 27 to 29 years of your life since it takes this long for Saturn to return to its starting position of when you were born. This is why many people have deep revelations and mid-life insights just before they turn 30 – it is the reality check from

Saturn! This planet rules Capricorn, which is why there is so much practicality embedded in the mind of a Capricorn. Find out where Saturn lands in your chart to understand your biggest lesson in this lifetime.

OUTER PLANETS

Moving onto the outer rings of planets now - Uranus, Neptune and Pluto are slow movers, often taking generations to complete one full orbit. This creates a societal or generational influence rather than one on an individual level.

Uranus

Aquarius is ruled by Uranus, and this is seen by the future-orientated aspects of this sign. Uranus moves beyond tradition. It does live on the outskirts of the solar system after all. Banished to the sidelines, this planet thrives in originality and progressivity. If you need to be innovative or want to seek enlightenment, you can take a look at Uranus in your birth chart. This will guide you to the place in your life that you can let your creativity flow. This planet also highlights intuitive abilities and seeks to investigate new topics and themes all the time.

As a slow-moving planet, its placement in your chart will be similar to those in your generation. This means that whatever house Uranus falls in, is one that will be a collective movement during your time. Think rebellions against established order, revolutions and a new order emerging.

Neptune

Also on the outskirts, we find the planet Neptune, the ruler of Pisces. Neptune is a planet of dreams, psychic abilities and transcending illusions and this is another slow-moving planet that influences generations at a time. The action of Neptune will be seen as a generation moves toward their version of an 'ideal world'. Neptune is receptive to new spiritual ideas, perhaps explaining why we have a new spiritual or religious revolution every couple of hundred years. The symbol for Neptune is a trident with three prongs but is commonly interpreted as an aerial for receiving divine guidance and inspiration. Wherever Neptune sits in your chart, will indicate where your generation is receiving guidance and how spiritual connection will look for your people.

Pluto

Pluto is the ruler of Scorpio, which explains why this sign isn't afraid to go out into the depths and darkness and be on the outside ring of society. Pluto is all about renewal and rebirth, particularly in a spiritual sense. Where Pluto lands in the chart is where a generation will see a deep search for truth and meaning. This will cause changes, power struggles and fight for control in these areas of life.

This area will also be a place where a generation seeks changes and transformation, eliminating the old ways to make space for the new!

There is also the association with Pluto and the underworld, making it important to see our darkness or destructiveness as humanity, and to seek places in which we can right our wrongs. But to be afraid of the dark is to neglect the lesson of Pluto completely. Embrace where this planet lands in your chart, and use it to your advantage.

Chiron

Chiron is not quite a planet, but rather a comet with an erratic orbit around the sun. It is commonly known in astrology as the 'wounded healer' for it points to the aspect in our chart that deep healing needs to take place. It represents our deepest wounds, not only in this lifetime but in previous lifetimes as well. But from our own wounds, emerge our strongest healing powers. Where Chiron lands in your chart, will indicate a place of talent, knowledge and natural abilities. This can be a great guide to us during our human experience, for we can remember that within the pain always lies a powerful gift.

Vedic Astrology

Within Vedic Astrology, there are three more 'planets' that are commonly referred to. These are the asteroids Ceres, Juno and Eros.

Ceres

Ceres rules food, nourishment, female cycles, motherhood and family relationships. Look to Ceres in your chart to find out how you can best nurture others. For example, if Ceres is in an Earth sign, you might nurture others with your cooking or physical touch. But if Ceres is in a Fire sign, you might be nurturing with your passionate support and encouragement of others.

Juno

Juno represents commitment and betrayal, and the giving and taking that occurs in relationships. You can look to where Juno sits in your chart to understand where you might be most triggered within a relationship. With this understanding you can work your triggers into your potential power, finding more balance within yourself.

Eros

Eros is the god of sexual love and desire. Where Eros lands in your chart will indicate the things that you find irresistible. With passion on the mind, Eros will show you where you feel most creative, joyful and turned on by life.

Conclusion

The zodiac signs are accentuated by the energy of their ruling planet. The planets highlight the meanings behind their placement on your chart. But to take it deeper, we need to see how you can use these planets within the 12 houses to really gain an understanding of the areas in your life which they influence.

You're on your way to becoming an astrologist yourself! Let's add another layer of information as we take it to the next chapter...

Chapter 5

The Twelve Astrological Houses

Now that you have learnt about each of the individual planets, you are going to learn about how each of these planets can manifest themselves into your life depending on which House they are in. There are 12 houses in Astrology and each house represents a different aspect of life. To understand this better, your natal chart is divided in different ways. There are hemispheres, quadrants and then houses. In this chapter we will explore what all of this means so that you can apply greater meaning to your own chart.

The difference between hemispheres and quadrants is that when I refer to a hemisphere, I am referring to one half of the chart, and when I refer to a quadrant, I am referring to one quarter of the chart.

HEMISPHERES

When you look at your chart, there are many ways to gain a broad perspective about personal characteristics and personality traits. One way to do this is through dividing your chart into Hemispheres. If you were to draw a line horizontally, you would find the north and south hemispheres. These are counter intuitive though, as the north is found at the bottom half of your chart and the south is the top half. Now, if you were to draw a line vertically, you would divide your chart into the eastern and western hemispheres. The east being on the left-hand side and the west being on the right-hand side.

When you look at the hemispheres you will begin to understand how your personality is shaped depending on if you have a cluster of planets in the north, east, south or west.

Let's see what each hemisphere represents...

Northern

The Northern hemisphere is made up of Houses 1 to 6 and is positioned as the lower half of your chart, below the horizon. These first six Houses are the ones that directly impact who we are as a person. They contribute to our personality and form our identity. If you find that a lot of planets fall in the Northern Hemisphere of your chart, you may find yourself more private than others. Holding your personality inwards and dealing with challenges more internally rather than externally.

Southern

The Southern hemisphere is above the horizon, in the upper half of your chart. This consists of Houses 7 to 12 and these represent the more social part of your chart. If you find that you have a lot of planets in the Southern hemisphere, it usually means you are more concerned with the outside world and may feel outgoing in social settings. You connect with others and project your problems outwardly to find solutions.

Eastern

The Eastern hemisphere is on the left side of your chart and includes the first three and last three Houses (Houses 1, 2, 3, 10, 11, 12). If you have a lot of planets in the east, you are most likely self-motivated and good at generating ideas and inspiration on your own. You may find that you are action-orientated and a big believer in free will. You go through life knowing that you are the creator of your own destiny.

Western

In the Western hemisphere, on the right side of your chart, you have Houses 4, 5, 6, 7, 8 and 9. These houses tend to concern the reactions of others, meaning that, if you have many planets on this side of your chart, you will be more likely to follow what others do rather than to initiate action. If your chart is heavy in the Western hemisphere, then you may also feel like you have a life path that you are following, without much choice as to where it will take you.

QUADRANTS

Aside from Hemispheres, the Natal Chart can also be divided up into four other sections, called Quadrants. If you were to draw two lines through your chart – one horizontally and one vertically (with the intersecting point being the centre of your chart), then you would see four quarters, also known as quadrants. These are simply named Quadrant I (Houses 1, 2, 3), Quadrant II (Houses 4, 5, 6), Quadrant III (Houses 7, 8, 9) and Quadrant IV (Houses 10, 11, 12).

The quadrants tell us about our developmental stages in life, and they can indicate at which life stage you will have the most challenges, depending on where the majority of the planets lay in your chart.

For example, if you have the sun in your first quadrant, it is like having your sun in houses 1, 2, and 3 all at once. The same goes for every planet, for example Mercury in the 4th quadrant is like having Mercury in houses 10, 11 and 12 together.

Some astrologers use the quadrant system to replace the 12 house system as it gives a broader overview of the meaning of the houses and can be a simple way to interpret a chart!

1st Quadrant

The first quadrant is about self-development and independence. It consists of the houses that are focused on gaining self-awareness and clarity about the ideas that make up our inner world. People with a lot of planets placed in this quadrant are usually very self-sufficient and focused on their own goals. The lesson here is to reach out to others and to learn how to collaborate when needed. If many planets are in this quadrant, you may find yourself asking – who really am I? And constantly seeking the answer through your life choices.

2nd Quadrant

The second quadrant is also about developing awareness, but this time it is focused on things outside of the 'self'. For example, the focus is on family, partnerships, friends, children and colleagues. Whatever planets land in this quadrant will indicate how you choose to interact with the world around you. People with a lot of planets in this second quadrant will have more complex interactions with the outside world. The lesson here is to create balance and unity across all relations. The life question you may find yourself asking is – what do I feel?

3rd Quadrant

The third quadrant is concerned with relationships, intimacy and life and death. It is a transformative place, for every relationship gives you new insights about who you are as a person. This creates a constant rebirth of you over and over again within a lifetime. The lesson here is to become your own person within your partnerships and to evolve through healthy cooperation with others. People with lots of planets in their third house will most often be asking – How do I bring myself back into balance?

4th Quadrant

The fourth quadrant is all about society. It has an emphasis on career, purpose, reputation, and public image. Planets based here are focused on finding out what humanity is here to do as a whole. People with a heavy base of planets in this quadrant may become too attached to their reputation in the world. The lesson, therefore, is to grow and learn through social contributions, and giving back to those in need. The question for people with a focus on planets in this quadrant will be– what do *I* believe?

What are 'Houses' in Astrology?

Your natal chart is divided up into 12 segments, like pieces of pie. You may have been wondering what these lines in your chart mean, and now you are about to find out. Each slice of pie represents an Astrological House, or one key area of life. All of the planets will be placed within the Houses of your chart and this will influence a certain aspect of your life. We will go through each of the houses in detail so you can start to interpret your own chart with this new knowledge. Don't worry if you are feeling confused, there are so many ways to divide up the chart that it can seem quite overwhelming! The good news is that the more ways you learn to divide your chart, the more insights you can gain!

Characteristics of Each House:

Let's take a look at what each house means.

1st House – House of Self

The First House is the one where your ascendent sign is at the time of your birth. It represents your unique self and so begins the journey through all 12 zodiac Houses, but also through the cycles of your life! It is naturally ruled by Aries and Mars. Being the House of Self, this one is concerned about self-image and who we are in the world.

It is all about new beginnings and how we make new impressions or take initiatives for new projects. If you have any planets in your first house, they will show up significantly in your personality. For example, if mercury is in your first house of self, chances are that you will mostly express yourself through verbal or written communication with others. Whereas, if you have Mars in your first house, you might often show up to others more confident and brave than you actually intend to!

2nd House – The House of Value

This house is related to all things tangible and of value. This is money, yes, but also earthly possessions and pleasures such as your five senses. To touch, taste, smell, see and hear are valuable but basic sensory experiences. Naturally ruled by Taurus and Venus, this explains the second house's need for physical comfort and security. Aside from valuing material things, this house also relates to how you value yourself. Self-worth and self-esteem issues will be highlighted in this House, particularly if you have a few planets residing here. Depending on what planets land in this House on your natal chart, you may be cruising through life with financial stability and ease, or working hard just to earn a living. It is written in the stars, but don't forget there are other aspects to consider too to make your life easier!

3rd House – The House of Communication

Ruled by Gemini and Mercury, the ones who are concerned with clear or chaotic communication (depending if Mercury is retrograde or not!) this house is where we move outside of our own inner bubble and start to interact with the outside world. If planet placements are smooth here, then you may be talking your way through life with ease. This house can also get messy if too much gossip or small talk is involved. Try to find out ways that you can communicate clearly with others, but first this means getting clear with yourself. Observe which sign rules this house in your chart to become clear on the best and most effective ways for you to communicate with others.

4th House – The House of Home

The house of Home is your foundational base. Ruled by Cancer and the Moon, this house is concerned with safety and security. It's literally found at the bottom of the Zodiac wheel, representing the base upon which you build your life. Physically when we think of a home, we tend to think of a safe and cozy place, but this House can also be about creating a home within yourself. Finding emotional comfort and solace in your own body as well as your surroundings. The Fourth House is also related to your family line and ancestors, those people that have been a part of your past and nostalgic memories. If there are a lot of planets placed here in your chart, you may have a lot of family lessons arising this lifetime.

5th House – The House of Pleasure and Creativity

This house is naturally ruled by Leo and the Sun, making it a warm and pleasurable place to be. This house is focused on fun and if there are big planet placements in this area of your chart, you may find that you like to focus on the exciting things in life. Activities that may be affected by this house are creative pursuits, hobbies, romance, love affairs, relationships and sex. Also, along the lines of creation, this house can represent your desire to have children. Leo and the Sun are heavily influenced by the heart, so whatever happens in this house must be done with a spark of love. Passion and pursuit are strong here, but it depends what planets you have in here to be able to discover what exactly you will be pursuing. If Venus or Mars are found in your 5th House, you might find yourself chasing relationships. Yet if Pisces lands in the same spot, you could be chasing your dreams.

6th House – The House of Health

Virgo, the practical earth mother, rules over the sixth House, accompanied by Mercury. This creates a healthy and steady place to focus on health and service. This house will give you insight as to how you make (or don't make) yourself of service to others. The aspect here in your chart will highlight your career goals and work ethics, with an underlying sense of your attitude to work.

Aside from work, the importance of health is also evident in this house. Nutrition choices, exercise routines and a healthy lifestyle will be hot topics. Are the planets in your chart helping or hindering you here? Look at which signs are ruling this house so that you can understand what qualities you need to nourish your body in the best ways.

7th House – The House of Partnerships

Laying directly opposite the House of Self (First House), this seventh house is concerned with your partnerships with others. This can be in a romantic sense, but also consider what the house means for your business partnerships. With Libra naturally ruling over this house, and Venus in tow, there will be a constant search for balance and equilibrium within partnerships here. In the end it depends what planets land in here that will determine how your relationships tend to fare. Look to this house when considering contracts, business deals or perhaps marriage, for it will highlight what you need to be aware of before entering these partnerships.

8th House – The House of Birth, Death, Sex and Transformation

This house is a mysterious one, naturally ruled by Scorpio and Pluto, there are many facets to its reputation. With life and death being at the forefront, the theme of rebirth is strong here, for it corresponds with the power of transformation. These are all themes that relate to a change or exchange in energy. Which allows sex to fit into this house as well, for sex is another exchange of energy that can lead to personal growth and healthy boundaries when with the right partner. There are often sacrifices and compromises that must be made in this house for balance to be restored, so take a look at the eighth House in your chart to find out which planets will be spicing up your life.

9th House – The House of Big Ideas

With Jupiter and Sagittarius ruling the ninth House, it is no wonder that big ideas are thrown into the mix. With philosophical thinking and a focus on chasing new knowledge, there is a true sense of adventure that resides in this house. Travel is a major topic here, as is education and religion. Look to which planets are in your ninth House for you to determine which type of adventure you will be on in this lifetime. If you've got your Jupiter in your ninth House you'll be on a constant search for new adventures, languages, flavors and cultures. Whereas If you have the moon in your 9th House, you'll have a deep need to explore the emotional aspects of life and ask the bigger questions about why we are here.

10th House – House of Career and Destiny

This house is concerned with your public image and your path in life. Located at the very top and middle of your chart, in a place also known as your 'midheaven', this house is showing you the steps toward your overall destiny. It can also represent authority or the relationship you have with a father figure in your life. If you want insights on your career path, this house is the place to look. It can also hint to you the ideal purpose for you in this lifetime. Moving upwards, it shows us what we are ambitious about and how to use this to enhance our professional reputation.

11th House – The House of Community and Friends

This house looks at the bigger circle of friends or community you have around you. Naturally ruled by Aquarius and Uranus, the eleventh House explores how you work together in teams, spread your network and look to collaborate to improve the state of humanity. Whether you are a leader (Mars in 11th House), or a follower (Venus in 11th House), you will understand more about your role in society when you look at this aspect of your chart. As you gain more insight about the community, you will start to shape your collective and long-term goals to make the world a better place. This house can also give hints as to how you work in group dynamics.

As well as friendships, this House is also related to technology and how you may best work with electronic devices to improve your life.

12th House – The House of Subconscious and Secrets

Over to the last house in the natal chart, we end with secrets. Ruled by Pisces and Neptune, this house holds a sense of mystery. Looking to this house can uncover some of your deepest wounds, and it can illuminate these areas in your life so that healing can occur. Ruling our unconscious mind, this house is interested in dreams, hypnotic states, sleep, karma and past lives. By doing a little digging and getting into the depths of this house, you will begin to see the true nature that you are beneath all of the human facades that we try to wear. Depending on what planets land in this part of your chart, will give you insights as to how secretive or non-secretive you like to be!

Meaning of Empty Houses

There are 12 Houses in Astrology but only 10 planets, which means that even if every planet lands in a different house, there will still be empty houses in your chart (houses without planets). This is natural and it simply means that you may not have as much focus on that area in your life as much as you would on a House with multiple planets.

CONCLUSION

The Houses are important to understand what aspects and areas of your life are being impacted by the planetary placements in your birth chart. Once you understand this, you are almost there! With the Houses, Planets and Zodiac signs it is time to start putting it all together so that you can read your own birth chart! Once you learn how to read charts, you can apply this wisdom to other peoples charts as well and excite everyone with your new knowledge!

Chapter 6

Putting It All Together

Now that you have the tools, it is time to put it all together and understand your birth chart! Once you learn how to read your chart, you can begin to decode the messages of the stars. Know that it is not easy to read a birth chart, and that's why professional astrologers take years of practice to become a master of their craft. But with a little practice and commitment, anyone can learn the skills needed to decipher a birth chart.

What is your Astrological Blueprint?

A blueprint is a plan, model or design that has technical aspects to it. In regard to your personal astrology chart, you have a unique blueprint that contains the essence of your soul's DNA. The reason that all of our charts are so different is because it depends on the exact time and exact location that we were born.

The location and time are significant because at the time you fully entered the Earth, the planets were somewhere very specific in the sky. Celestial bodies are constantly moving, so you have to imagine that we pause the skies at the time of your birth and take a snapshot of what is happening up there.

This Astrological chart, commonly known as your birth chart or natal chart, literally holds the energy of your entire lifetime. It is through discovering the meaning of this chart that you can understand who you are at your core. Your birth chart is so individual that it could even be likened to a passport, for it holds valuable information about your identity.

As you learn to read your birth chart, you will discover that the placement of the planets really does define who you are. It allows you to tell a story of your personality, goals, drivers and challenges in this lifetime. So, let's take a look at the basics.

WHY IS YOUR BIRTH TIME IMPORTANT TO KNOW FOR YOUR BIRTH CHART?

As the planets and celestial bodies are always moving, the exact time that they will be charted on your personal birth chart will differ depending on the hour and the minutes. The exact time of your birth determines what your rising sign (Ascendant) will be, for it is the sign that was on the eastern horizon at the time of your birth. The sign changes around every two hours which means that if your birth time is calculated wrong, you will have the wrong rising sign or in metaphorical terms – the wrong personality!

As your chart also gives you insights about your health, career and destiny, it is not something that you want to misinterpret or you may be taking the wrong advice!

WHAT DO YOU DO IF YOU DON'T KNOW YOUR BIRTH TIME?

Not knowing your birth time is a common thing, and often people only start asking this question when they begin their astrological journey.

If you have asked your parents about your birth time and they can't remember, or if you were adopted and the information of your birth time is no longer available, you do not need to worry, for there are other ways to find out your birth time!

Please note, if someone you know simply guesses your birth time, it is also good to double check, for calculating your birth chart with the wrong time will give false results throughout your whole birth chart.

5 WAYS TO FIND OUT YOUR BIRTH TIME!

1. Check your birth certificate!

If you don't have one or your birth time is not written on your certificate, you can try calling the statistics registry office where you were born. You may even try calling the hospital that you were born at or perhaps the church you were Baptized at to request more information on your records.

2. Look through your baby books or photos.

This may give you insight as to what time of day it was at the time of your birth. You can also look to see if there are any date or time stamps on the photos taken after your birth or information written in your baby books.

3. Ask parents or family members to confirm.

After looking at some of the baby photos or books, talk to family members to ask if they can confirm the results that you have found. Anyone that was at the birth may remember specific details to confirm what time of day it was.

4. Use muscle testing or a pendulum.

You can use these things to test the true answer of your body. If you are unfamiliar with them, you may seek out a qualified kinesiologist or psychic therapist to work with.

5. Go to an astrologer that specializes in Birth Time Rectification

If you still haven't gotten the answer, seek out an astrologist that specializes in Birth Time Rectification. They will be able to work it out through putting together a few sample charts and determining which one fits you best!

How do I figure out what my chart says about me?

Once you have your birth time, date and location, you can put these details into an online calculator to view your chart! There are many free resources available online. Once you have an image of your chart, you can start to apply the knowledge from this book into the aspects of your chart!

I recommend starting a small notebook or journal in which you can record your findings. Keeping your astrological insights in a clear and structured method will help you to create a bigger picture of your personality over time.

Of course, you can seek the help of a professional astrologer for interpretation of your chart, but if you do this, make sure you find a reputable one with good reviews and feedback.

How to read a Birth Chart – The Basics

Looking at your chart, you will see a circle with lots of lines and symbols covering it. At first glance, this can look very confusing! But, we are going to break it down into easy and digestible steps.

The outside ring has the symbols of the 12 zodiac signs, and this can help you to work out which sign is ruling which house. The signs are always in the same order, but they will shift around the chart depending on what your Ascendent sign is.

The inside ring is also split into 12 sections, and these are the 12 Houses. These 12 Houses are marked by roman numerals so that you can work out which house is which. The first house begins on the left side and then they increase anti-clockwise, creating a full circle. Not all houses are made equal, and they may vary in size on your chart, making some houses bigger (more degrees wider) than others.

Between the inner and outer rings, in the body of the circle you will see the symbols that represent each of the planets. By looking at where the planets are placed you will be able to decipher which house each one lands in.

When you compare the sections of the Houses with the sections of the Zodiac Signs, you will be able to work out which sign (or signs) are ruling each House.

Now, to begin with your personal interpretation....

THE PRIMAL TRIAD

You can begin by finding the Primal Triad. This is the Sun, the Moon and your Ascendant (or rising sign). These are usually referred to as the three most important aspects in your chart, for they make big bold statements about your personality.

Your sun sign (indicated by a small symbol of a sun) represents the core image of who you are. You can see what house and sign this falls into for more detailed information about what aspects of your life the sun is illuminating for you this lifetime.

Your moon sign (indicated by a small crescent moon) represents your emotional self. It is how you feel in yourself and your surroundings. It represents the way you process your emotions and so seeing where this lands in your chart will indicate the most emotional aspects of your life.

Your ascendant is your mask to the world. It is indicated on your chart by the letters AS and it is where your first house begins. Your ascendant sign represents how you show yourself to other people. See what sign is ruling your ascendant to understand how other people perceive you.

Now, once you have understood the basic zodiac signs that rule your sun, moon and ascendant, you can start to add more layers to your astrological personality. This will allow you to look at the bigger picture.

Combining Planets, Houses and Signs

- The Planets represent mental functions and ways of thinking.
- The Houses represent areas of life.
- The signs represent characteristics and personalities.

When you put these three aspects together, you can really start to add a personal touch to your chart.

First of all, choose one planet. The planet represents the 'what'.

Then, look at what House the planet has landed in. This will show you what area of your life this planet is affecting – it represents the 'how'.

Then, see what Zodiac sign is ruling this area of the chart, for that will represent the 'why'.

For example, you might ask a question like:

How can I communicate more clearly?

For this, you would look to Mercury, the planet of communication (what), then you would see that it is in the fifth house of creativity (how) and that it is being ruled by Gemini in your chart (why).

> **What:** Communication style
> **How:** With pleasure and creativity
> **Why:** Gemini is the expressive and curious personality

Someone with this placement would be a creative and curious story teller, who enjoys captivating an audience with their expressive recounts of their life.

Now it is your turn to create these personality aspects of the placement of planets in your chart.

Use the following guide to put together a story about each planet.

Planets:

- **Sun** – your self
- **Moon** – your emotions
- **Mercury** – your communication
- **Venus** – your love
- **Mars** – your action
- **Jupiter** – your place to expand
- **Saturn** – your boundaries and lessons
- **Uranus** – your generations future
- **Neptune** – the goals and psychic undercurrent of your generation
- **Pluto** – renewal and rebirth of a generation

Houses:

- **1st House** – Self
- **2nd House** – Value
- **3rd House** – Communication
- **4th House** – Home
- **5th House** – Pleasure and Creativity
- **6th House** – Health
- **7th House** – Partnerships
- **8th House** – Transformation
- **9th House** – Big Ideas
- **10th house** – Destiny
- **11th House** – Community and Friends

- **12th House** – Subconscious

Zodiac Signs:

- **Aries** – Fiery and action orientated
- **Taurus** – Grounded and stable
- **Gemini** – Expressive and curious
- **Cancer** – Introverted and Emotional
- **Leo** – Bold and Outgoing
- **Virgo** – Practical and patient
- **Libra** – Balance and justice
- **Scorpio** – Mysterious and truth seeking
- **Sagittarius** – Adventurous and unafraid
- **Capricorn** – Resilient and Determined
- **Aquarius** – Innovative and individualistic
- **Pisces** – Intuitive and sensitive

Now, these are just very brief descriptions for the planets, houses and signs. But they will be enough for putting together the basic descriptions that you need. With this simplicity we can create an example that will help you to understand how to read your chart.

Note: For more depth in your reading, you can go back to the previous chapters to revise the descriptions of the planet, house or sign and then add more details to your story.

For example, if you have the planet Venus, in the 4th House, ruled by the sign Libra, then this means your way of loving (Venus) will be in the form of creating a stable home with your partner (4th House) and you will seek balance and justice within your relationships (Libra).

Or, you may have the planet Mercury, in the 9th House, ruled by the sign Sagittarius.

This could mean that you love to communicate (Mercury) your big ideas and plans (9th House) with fearlessness and a sense of adventure (Sagittarius).

Now, go ahead and look at your chart. List the planets and then next to each one write down the House and the Sign that they are in. From here, see if you can create one sentence for each planet, just like the example above, to start giving your planets your own personal characteristics.

Even if you feel that some of these characteristics don't fit your life right now, it could be an issue that you worked on in the past or will work toward in the future, so stay open-minded to what insights might occur.

More Depth and Detail:

Once you have understood the basic placement of planets and what they mean for you, you may choose to add even more detail. This can be done by looking at the following aspects:

The Nodal Axis – Nodes of the Moon

The two lunar nodes represent your future and your past. On the chart, these are two mathematical points that are calculated in relation to the moon and its orbit.

The south node tells you where you've been and the north node tells you where you are going. They stretch in opposite directions, pushing you to become the complete opposite of who you used to be.

Note: some natal charts will only show the north node, but you can always know that your south node is directly opposite (180 degrees).

South Node

Our south node represents where we are coming from and what we are bringing with us into this life. It is the accumulation of events and skills, qualities and gifts that we have collected from past lives. They are the aspects of ourselves that we have already mastered, and so we bring them forth into our current life so that we can continue to progress forwards. From the South, we grow up and into the North.

North Node

The North node is like the light at the end of the tunnel. It is the path that we are walking towards total mastery and self-discovery. Some astrologers even go as far as to say it is our destiny, the place that we will ultimately end up in, no matter what path we choose to get there. Following the steps toward the north node is not always easy, as big lessons are bound to occur on the way. But, if we do keep the end goal in mind, we are sure to feel fulfilled as we work our way through the challenges.

Ascendant

The ascendant is on the cusp of the first house, defining how we experience our first impressions of life. You will find it on your chart by the letters AS or ASC. Depending on what sign the ascendant falls into, will determine how we show our physical appearance and body ego to the world around us. It also represents our natural defense mechanisms, and the tactics that we put up when we feel threatened. How we use our ascendant sign will determine how much we let others into our lives. If you have an open and honest ascendant sign you will be able to forge friendships and partnerships easier than if your ascendant falls into a deeper and more reserved sign. Allow yourself to get to know the attitude of your ascendant well so that you can choose which aspects of it you like to embody and those that you would like to tone down a little.

Midheaven

Midheaven is represented by the letters MC (meaning *Medium Coeli* - middle of the sky) on your natal chart. It is on the cusp of the 10th House, making it the point that most often lands at the very top of our charts. Being at the top, in the spotlight, the MC is seen as the most public point in our charts. It can relate to our life path, especially in regards to our social standing and community reputation. Depending on which sign rules the MC, will determine how you choose to be seen in public. For example, if Leo is ruling the MC, you will be loving the Limelight, but if Cancer is the one at the top of your chart, then you are more likely to be reserved in public places.

To make it clear, the midheaven will represent your most visible achievements, material accomplishments and overall responsibility in the world or society. It is closely related to your public face, the one you show to complete strangers (rather than the more personal face that is shown by the ascendant).

Fortuna

Fortuna goes by a few names: Part of Fortune or Arabic Part, both indicating sensitive points in your chart that use a special formula. This formula is:

Fortuna = Ascendant + Moon – Sun (for day charts)
Or
Fortuna = Ascendant + Sun – Moon (for night charts)

To determine if your chart is a day or night chart, you can look at the Sun's position. If the Sun is above the horizon (Houses 7 to 12) then you have a day chart. But if the Sun is below the horizon (Houses 1 – 6) then you have a night chart.

This will give you the Part of Fortune, a placement in your chart that represents success in both career and health pursuits. Given the name 'fortune', this point in your chart will highlight your innate abilities and qualities that are natural talents. The natural expression of things that you are good at will ultimately lead you to prosperity.

Often, Fortuna is not found on birth charts, but to calculate this point, you can work out the distance in longitude that the Moon lies from the Sun. Then calculate this same distance from the Ascendant and you will find the place of Fortuna.

Vertex

Vertex is another little known point in astrology, but one that represents karmic or fated connection. So, really it is worth looking at. This placement is said to be activated at certain turning points in our life. These will be big events such as meeting a partner or ending a relationship. Some even say it is activated when there is a birth or death of a loved one.

When you are in a serious relationship, you can compare both of your charts to see where your Vertex points might connect or intersect, defining the type of love within the relationship. If your vertex point correlates with your partner's ascendant or other significant planet, then it can define how your relationship unfolds.

Sun Moon Midpoint

As the name suggests, the Sun Moon Midpoint is exactly halfway between the Sun and the Moon. Like the longing of connection between the sun and moon, they meet in the middle to create a harmonious life on earth. This is also similar to what the midpoint represents – the desire for companionship. It can indicate a need for external companionship with others, or even a lust for internal companionship within the self.

Retrograde Planets in your chart

When planets go retrograde, it means that they appear to move backwards in the sky. This causes chaos and confusion. If you see a little 'R' next to a planet on your birth chart, it indicates that those celestial bodies were in retrograde on the day you were born. When a planet is in retrograde, the energy of it is somewhat blocked, meaning that those areas of your life are not as easily expressed. So if you do see a retrograde symbol next to a planet, just know that you are going to be working a lot with this planet to gain deeper insights into yourself. For example:

- **Mercury Retrograde** – means you'll be working on thinking and communication this lifetime.
- **Venus Retrograde** – indicates your tough lessons will be in love, beauty, pleasure and creativity.
- **Mars Retrograde** – this means you'll be fighting your energy levels and ambition back into balance regularly.
- **Jupiter Retrograde** – this will give you extra work around abundance and good luck.
- **Saturn Retrograde** – will be asking you to mature and take responsibility for your life.
- **Uranus Retrograde** – is a challenge to express your authenticity and originality.
- **Neptune Retrograde** – a chance to seek spirituality and compassion in all aspects of your life.

- **Pluto Retrograde** – you'll be constantly challenged by your shadows and pushed toward transformation.

Intercepted Signs and Planets

The Houses in astrology vary in size, and this causes a different number of signs to fit within one House. An interception can take place when one house is greater than 30 degrees and engulfs one entire sign. The result of this is that one sign is totally enclosed in a house. Whichever sign this happens to, it must also be true for the opposite sign on the other side of the chart. For example, if the 4th House is intercepted, so too is the 10th House.

If there is a planet within the intercepted sign, then these are also considered intercepted planets.

The significance of interception is that you will most likely have a hard time expressing the qualities of that sign in your life. It often makes you bottle up or hide these certain qualities in yourself, ultimately creating shadows in these parts of your life.

There can be both challenging and positive aspects of intercepted planets. Here are some examples of what it might mean for you., if you find an interception in your chart.

Intercepted Sun means that you may not feel as radiant and powerful as the sun itself, but with a little push from the outside world or your loved ones you will still be able to reach your goals.

Intercepted Moon can mean that your emotional stability and moods are more closely related to intuition compared to others.

If Mercury is intercepted, you might also be more reflective rather than quick to talk, letting yourself find the right way slowly but surely.

Intercepted Mars may surprise you when you take the lead or get good results in a natural way.

Jupiter being intercepted may mean that you never let anyone tell you that your dreams won't come true. Your positive spirit remains strong.

If Saturn is intercepted, you may need to overcome a sensitivity to criticism so that you can actually use it in a helpful way.

Uranus intercepted means that you are an independent spirit and your ability to assess the future will set you free.

Neptune intercepted probably means that you like keeping secrets right up until the moment where all will be revealed.

Finally, Pluto being intercepted means that you may be challenged to make difficult choices that scare you. But after all, it is for your own growth and transformation!

Placement of the Planets

You can also refer back to Chapter 5 to understand the hemispheres and quadrants of the chart and what that means for you!

CONCLUSION:

Now that you have explored a few areas of your chart, you should have more understanding about your true nature, your strengths and weaknesses, your destiny, and your history. You can refer back to your chart at various points in your life to gain a greater insight as to how some of these aspects are either currently showing up or perhaps they already did play out in your life. For example, look at where Mercury sits in your chart during a Mercury retrograde period, or study your placement of Saturn when you are around 27 to 29 years old to learn about what your Saturn return is teaching you.

When you feel comfortable reading your own chart, you might even like to have a look at other peoples to learn more about what parts of their personality that you are compatible with. As you uncover some of the aspects of your own chart, you will also discover what you seek in a partner. That is why astrology is often used to create the perfect love story. You can try to put this together in the next chapter!

Chapter 7

Finding Your Perfect Partner

There is so much that you can take away from your chart including your strengths, weaknesses, personality traits and ways of living, but how can you use that information to find the perfect lover? Well, you are in the right place, because this chapter is about to dive into the juicy world of relationships and astrology.

We all know that as much as we search and look for the perfect qualities in a partner, simply not everyone can be right for us. This is because we are all unique individuals. But as you may have noticed through reading your chart, there are some parts of it that suggest you seek a reliable partner or an adventurous one. Perhaps some aspects of your chart want you to find someone who is creative, or someone who is practical. This is where we can look to astrology to find who the perfect match for you might be!

Just one note before you read too deeply into this – there are many ways to interpret the signs, and this is simply a fun way of looking at traits and qualities in certain people. Please, don't go breaking up with a long-term partner after reading that your signs are not compatible, for there is much more to it than just the sun signs. Sometimes, challenging partnerships are also needed for they help us to expand and grow into better and brighter versions of ourselves. So, with that out of the way, let's have some fun.

Due to online dating, and being more connected than ever, we are meeting people that 100 years ago it would have been impossible to meet! This makes dating challenging, because rather than having 5 nice potential partners to choose from, suddenly you have 5000!

Now, of course there are benefits to having access to so many people, for it means that we can connect, meet and be inspired by others almost constantly. Whether you are looking for a man, woman, person of neutral gender or someone else on the LGBTQI+ scale, it is possible to find a plethora of potential partners. To narrow down the selection, you might like to start asking 'what's your sign?'. This question will surely break the ice on the first date, but it will also give you an insight into some of their favourable or not so favourable personality traits that you can expect to see as your future dates unfold.

There is more than one way to explore who is right for you.

One easy way is to consider the elements. For example, a Scorpio will be easily compatible with other water signs, for they all enjoy being in emotional depths together. Aries will also thrive off other fire signs, and they will take incredible and spontaneous adventures together. But although this might be great in the short term, you need to consider the long-term consequences too. For example, A Gemini paired with a Libra will have wonderful ideas and could be chatting for hours. But perhaps, what these air signs really need is an earth element such as Virgo or Capricorn to actually help them put their plans into action!

Same goes for other elements, for example an earthy Taurus probably needs a bit of Leo fire in their life to push them out of their comfort zone. For a breakdown of all of the elements, go back to Chapter 2 – and have a think about what element you may need in your life and why!

Now after you are familiar with the elements, you can use this information below to consider who might be your best match, which we will dive into soon.

But first I want to touch on the planets Venus and Mars.

Venus is the planet that tells us how we approach romantic relationships. Depending on what sign and house Venus is in on your chart, this will give a clue as to if you are a slow sensual, home-bodied lover, or if you like to express your relationships loud and proud like a Leo lion. Then, looking to Mars will also give clues about sexual compatibility, for the sign that Mars is in can tell us how one likes to initiate sex. For example, a fiery Sagittarius wants a quick sex session in the back of your car, but an earthy Virgo would prefer a cozy bedroom to make love in.

Love is exciting, so let's break it down into each sign.

Which Signs are you Compatible with?

This list will go through the top three matches for your sign! You can read these based on your sun sign and your moon sign.

Your sun sign reading will indicate whose personality matches yours the best.

Your moon sign reading will give you a clue as to who is most emotionally suited to you.

Aries

Aries are the first and most fiery sign of the Zodiac. With big appetites for desire and always wanting to be the best, it is good when these signs find a highly energetic partner. On a good day, an Aries is like a brave warrior or a fearless goddess, ready to take on the world, and they seek a partner to stand by their side and chase the sunsets with them. For this reason, the other fire signs are a great match for Aries. Sagittarius and Leo will invoke more fire, passion and heated love when they are paired with an Aries.

Whoever chooses an Aries as a lover must be brave and confident, and unafraid to stand their own ground. Libras also make a good choice here for they can seek justice and balance to bring out both partners best qualities.

Aries can be loud and expressive, so the more sensitive signs – looking at you cancer, Pisces and Capricorn - may feel overwhelmed with an Aries partner. Aries doesn't like their impulsive nature to be dampened, so they probably wouldn't choose these signs anyway, even though a little slowing down would be good for them.

Top 3 signs for an Aries to date:

1. Sagittarius
2. Leo

3. Libra

Taurus

Taurus is the hopeless romantic, the one that creates a cozy home base and seeks commitment. These earth signs are in touch with the feelings of their bodies, and they love physical attention and touch.

Taurus looks for stability in all areas in life, including in love, so already you can cross Sagittarius, Gemini and Aries off the potential partner list. Too much fire will dry out a Taurus's grounded nature, instead, people with this sign should seek those who also value stability. This makes a Taurus and Taurus couple the perfect pairing, because they both know how to look after themselves and also their partners in a way that they both feel seen and heard.

The taurus man is a strong and reliable person, while a taurus women is a practical and resourceful earth goddess. Matched with one of these, regardless of gender, you are sure to be spending time in nature, slowing down your life and enjoying delicious meals together. Scorpio is another great potential match for Taurus as this mysterious sign enjoys the dedication that they receive from their Taurus lover. Scorpio and Taurus are opposite signs on the zodiac wheel, and although this can create tension due to many differences, it is also true that opposites attract.

As for third best sign, Capricorn a fellow earth sign, and Pisces, a sensitive water sign rank as equal third! Capricorn shares the earthy values of Taurus, and Pisces encourages taurus to express their artform and creativity. Both creating stable partners for the loving bull.

Top 3 signs for a Taurus to date:

1. Taurus
2. Scorpio
3. Pisces or Capricorn

Gemini

Gemini is a sign that is current, eloquent and classy. They love to chat and share their insights about the world with others, but who really is compatible with this flirty sign?

Earth signs are a big no-go, for Gemini is flying up in the clouds, and any earth sign would bring them down. They don't need that practicality. A Gemini also can be flaky, and change their mind quickly, which means that they need someone who can keep up with their mental hamster wheel. This makes fiery Aries a good match, as an Aries is just as impulsive as a Gemini, and many crazy last-minute adventures will be had together. They are both fearless and will encourage one another to push their boundaries.

Fellow air sign Libra is also a good match for Gemini, as while Gemini swings from one opinion to the other, trying to work out what they actually think, Libra is there standing in the middle, holding a place of balance for them both. These are also two very talkative signs, so Gemini and Libra will be chatting until early hours in the morning.

But the number one match for Gemini is a Pisces. Both being mutable signs, they know how to go with the flow. Pisces likes depth and Gemini likes being curious and inquisitive, so together this pair is bound to uncover mysteries and connect like no others.

Top 3 signs for a Gemini to date:

1. Pisces
2. Libra
3. Aries

Cancer

Cancers were born to care, and they will look after anyone who allows it. It is their love language to be prepared with a solution for any potential problem. This sensitive sign knows how to listen and how to make you feel understood. This is why, the independent signs such as Aquarius, Aries, Sagittarius or Gemini, are not going to work!

Instead, Cancer needs someone who will be present with them, with strong communication skills and a sense of reliability. This makes Scorpio a number one match, for Scorpio is good at showing commitment to their lovers, and although they try to act cool and independent, they actually don't mind when their lover gets a little clingy. It shows how desired and wanted they are, and this fuels the relationship.

Another great pick for Cancer is the water sign Pisces. Pisces people often get lost in relationships and they tend to seek out someone who will care for them. Paired with a Cancer, the care will be given and well received. Both of these signs need to be nurtured and they are the kind of match that will be writing books and writing songs about their love.

If the Cancer is seeking a little more practicality in their life, they might also choose a Virgo. These earth signs will keep the Cancer balanced in the role of giving and receiving. These two signs love to give, and they love to have a cozy home. If this pair invites you for a dinner party, you won't want to miss it.

Top 3 signs for a Cancer to date:

1. Scorpio
2. Pisces
3. Virgo

Leo

The lion has a big heart, and although they enjoy being the center of attention, they also love it when their partner meets them in the spotlight! If you are dating a Leo, you will have to be comfortable putting yourself out there in the crowds. This is why Scorpio is the worst match for they like to keep to themselves and embrace their mysterious side.

But moving onto who would be a good match, we have to look to the other fire signs, for they also like to be seen! Aries and Sagittarius make a great match for a Leo because they can both keep up with the fast pace and passionate love. Leo is playful and so are the other fire signs, so a match here would be a delight for the inner child of both partners!

But one more pair that works well is Gemini and Leo. The airy Gemini has enough oomph to take on the boldness of the lion, and Leo likes to be flirted with. Gazing into one another's eyes, this match will know it is a pair from the stars! Both are adventurous and open to try anything, this makes the early dating days an interesting time for them!

Top 3 signs for a Leo to date:

1. Sagittarius
2. Aries

3. Gemini

Virgo

A Virgo needs someone who is caring, compassionate, curious and lighthearted. They are the keepers of the earth and order and they like to explore new ways of living that are comfortable and close to nature. They are very analytical, so they will quiz you on your ability to be realistic before choosing you as a partner.

Say goodbye to fire signs, they don't have nearly enough practicality to make a Virgo happy.

But this is why other earth signs make a wonderful match. Virgo and Taurus will make each other feel valued and loved. They will take care of each other, and they will be a pair until the day they die. This unbreakable bond thrives on their joint love for home, routine, and practical plans.

Capricorn also fits in well with a Virgo, for they too know how to prepare a 30-year plan! Virgo has got plans, and Capricorn has got efficiency, so these two will always get the job done, and they will feel the electrical love in their veins as they work on projects together.

Aside from earth signs, a Virgo is also a sweet match for a Cancer. Virgo can get anxious and worked up about everything being in order, but Cancer can teach them how to relax into their emotional body and take some time away from their minds. Cancers like to feel, and this can be a benefit for the overly analytical Virgo as they are taught how to find their intuitive mind. In turn, Virgos give back with their attention to detail and joy of reminiscing on past moments together.

Top 3 signs for a Virgo to date:

1. Capricorn
2. Cancer
3. Taurus

Libra

The keepers and restorers of balance like to make sure that their own relationships are balanced too. They want lighthearted fun, laughter and good memories made. In this sense, Scorpio and Virgo are not on the list for Libras as they are just too intense and stuck in their ways.

For fun, a Libra should choose to pair with an Aquarius, for cool conversations, and feeling like they've met their soul mate. The similarities of these signs are obvious and intense, but the problem is that both signs like to feel free. This could lead to a Libra and Aquarius partnership ending as quickly as it started, but there will be no regrets of the quality time spent together.

A Taurus or Capricorn pairing is also nice for Libra as these grounded earth signs aren't as analytical as Virgo, but they bring a healthy sense of commitment and stability to match Libra's need for balance.

The final pairing work noting is that of Gemini and Libra. The two minds meet, and this allows for complete and transparent authenticity. Both signs feel at home in this kind of partnership, and they will often be happy to cancel plans with other people, to spend quality time together. Whenever one starts a conversation, it will continue until the next day, or week, or even years down the track. These two will never get bored.

Top 3 signs for a Libra to date:

1. Gemini
2. Taurus
3. Capricorn

Scorpio

Who can live up to the intensity of Scorpio? Well, this is a question that a lot of Scorpios ask themselves. With their mysterious side, it can take a while for a Scorpio to let you into the true colors of their world, but once you are in, you will be protected and surprised by the authentic personality that you find within! With a tendency to back away and hide when things get overwhelming, a Scorpio is well paired with someone who can see this trait and bring them out of this rut. Aquarius makes a perfect match in this sense, for they are great and spotting things before they happen, but also finding innovative ways to get a Scorpio to talk rather than to hide!

Enjoying being in the depths, Scorpio is also well paired with Cancer, as this is another crustacean that feels deeply and isn't afraid of finding the truth. These signs will both be able to express their sensitivity freely without feeling judged by it, and they will never feel the need to hide. Cancers value security in relationships, just like Scorpio, so these two signs will be reliable in their partnership.

One more great match for a Scorpio, is another Scorpio, for these two have a connection like no others. Playing in the psychic realms, it's possible for these two to read each other's minds and communicate without using words. They will feel intuitively what their partner is thinking, and therefore emotions could be heated, but that's only because both partners are working on their jealously issues. As two mystery beings coming together, they create their own fire and stoke their own flames. They could eye gaze for hours on end, finding more depth in their partner each time.

Scorpios don't need to waste their time with a fire sign, for Scorpios like to hide in the darkness, while fire signs like to be in the bright light. Making both partners feel uncomfortable in these situations.

Top 3 signs for a Scorpio to date:

1. Scorpio
2. Cancer
3. Aquarius

Sagittarius

Fire is a strong element, and it takes a courageous soul to be with a Sagittarius. If you choose a Sag as your partner, be ready for exploring new corners of the world, eating out at midnight, taking spontaneous visits to friends and doing all of this without plans.

This means, Virgo, Capricorn and Taurus you are off the list, because let's face it, you won't try anything if it wasn't' written in your diary three months ago.

But Fire and Air make a good combination, as these two elements spar each other on, creating bigger and hotter wildfires and blazing a trail wherever they go. So, hello Aries. Combine Sag and Aries and you've got a potent combination. Two people unafraid to face the world, to leave their loved one's behind for a little time and to return from their journey triumphant. This combination is energetic, fun and will be guaranteed to be on the invite list to everyone's parties!

Pair a Sag with a Gemini though and it's a little more calm and collected in the way they take their dose of adventure. This combination will talk for hours, creating elaborate plans for where to go... but then burning the list and just jumping on the plane. The curious Gemini will keep the mind of the Sagittarius intrigued, and vice versa. There is no boredom in this partnership, for they both have an unlimited stack of ideas.

Another great combination is Sagittarius with Aquarius, for these two are easy going signs and accepting of each other's need for freedom. They understand the need to roam independently, but they also can learn how to roam together, for they have similar values and outlooks on life. Conflict will be resolved easily in a Sag and Aquarius pairing, for they are both signs that just want to move on and enjoy the best of life.

Top 3 signs for a Sagittarius to date:

1. Aries
2. Gemini
3. Aquarius

Capricorn

Capricorns don't open up to everyone, and you really have to win their trust before you enter into a relationship with this sign. Responsibility is high on the list for a Capricorn, so you need to be able to be held accountable for your actions and have a strong work ethic to be able to uphold their expectations. As well as this, Capricorn is also looking for loyalty, for they know that whoever they choose as a partner, they will be deeply loyal towards forever, even if conflict arises.

This means that earth signs are highly compatible, especially Virgo. Virgo is both passionate, caring and yet pragmatic, making it the perfect match for considerate Capricorn. These two will be able to create long term plans together, and work side by side to both feel deeply and express themselves to the world around them. They don't like the flashy fire signs getting in their way, but they also don't mind if the attention falls on them, for a Virgo and Capricorn pair will make each other feel comfortable in the spotlight.

Capricorn can also pair with another Capricorn, and this is sure to bring an abundance of appreciation, and romance within the relationship. A double Capricorn combination can butt heads at times, for they may want to compete their way to the top of the mountain, but usually once they have realized their conflicts, they will work through it together, reaching their goals at the same time and supporting each other all the way.

The other earth sign also works as a match for Capricorn. Taurus and Capricorn are both loyal, caring and like to create stability and security in their lives. Together they will create a routine that accommodates both of their needs, and they will strive for comfort and joy within their love life. They may come to stagnancy if they stay in their stubbornness for too long, but if they can move past this challenge, they will be a power couple.

In short, Capricorn feels most comfortable around its own earth element.

Top 3 signs for a Capricorn to date:

1. Virgo
2. Capricorn
3. Taurus

Aquarius

Probably the most unconventional of all of the signs, Aquarius can be a tough one to pin down in love. They are the sign that craves freedom, technological advancements and a need for deep introspective periods. This can make them both an idealistic figure in theory, but a hard one to keep up with physically.

Aquarius has many opinions, and likes to share their thoughts, so a partner needs to be both a good listener, but also one who stimulates interesting conversations too. This is where Gemini becomes a great partner. The two air signs could talk forever, feeding off one another's' analytical and curious minds. They both have eclectic interests, and it is okay if they hand out in difference circles, for then it also gives these two much needed time apart. Both Gemini and Aquarius are complex in nature, creating a bond over their need for self-expression.

Now, if you pair an Aquarius with another Aquarius, you'll be seeing a couple from the future. They have a humanitarian outlook on life and a deep need to find solutions to help humanity. This power combination will be savings lives somewhere in the world, deep in the work that is needed for human consciousness to evolve! If both partners can find their own individual flow within the partnership, then an Aquarian pair is sure to thrive.

While we are matching air signs, we may as well throw in Libra too. Libra will keep an Aquarius well-balanced and will remind them to come back down to earth occasionally. With both of their love languages being quality conversations, this pair is set for a life of curiosity and joy!

Top 3 signs for an Aquarius to date:

1. Gemini

2. Aquarius
3. Libra

Pisces

Last in the zodiac, and coming to an end as a wise, watery sign, these complex individuals need a partner who is unafraid to swim deep with them. Looking to fellow water signs, Scorpio can be a great match, as the emotional intelligence of these signs are both well developed. This can lead to deep conversations, speaking the truth and removing masks within the relationship. In fact, these two signs may become different people in their relationship as they drop their energetic boundaries when they are with the one they trust the most.

Surprisingly two Pisces are not a good match, as the two fish may find themselves swimming away from each other with an overload of sensitivity. Instead, a Pisces should find a Cancer or Taurus that is more emotionally grounded and able to hold the sensitive Pisces in their warm embrace.

Cancer has a tough shell, like Scorpio, they can deal well with sensitive people, while also holding space for transformation to occur. The kind and gentle presence of a Cancer and Pisces relationship will allow both partners to open up to each other and work on their feelings together.

When it comes to Taurus, the earth energy here will be a stable base for a Pisces to rest upon. It can be exhausting always swimming in the rough ocean waters, and Taurus is like a calm stable mountain to counteract the emotional variability. Together, this pair will be honest about their intentions and clear about their desire for long term love.

Top 3 signs for a Pisces to date:

1. Scorpio
2. Cancer
3. Taurus

REASONS YOU SHOULD DATE YOUR OPPOSITE SIGN

Everyone knows the saying 'opposites attract', just like magnets, and this is just as true in astrology. Although dating your opposite sign can sometimes feel like chaos, there are many benefits to pairing up with someone completely different to you!

In astrology, opposite signs are those that sit directly opposite each other on the elliptic (circle of your birth chart) and these are called polarities.

The polarities are:

- Aries + Libra

- Taurus + Scorpio
- Gemini + Sagittarius
- Cancer + Capricorn
- Leo + Aquarius
- Virgo + Pisces

Opposite signs are always different elements, but they are two elements that work together.

Fire and air when paired together make a bigger and brighter flame. Earth and water are needed together to grow a seed into a plant. These combinations are found in nature and have a beneficial effect, usually!

Opposite signs always share the same 'quality' – that's the cardinal, fixed or mutable qualities we have spoken about. So, lets see how each combination stacks up.

Aries + Libra:

This cardinal combination both like to be first at things. For Aries, that's the first to shake things up, whereas for Libra, it's the first to hide at the sight of conflict! Now, if well balanced, the Aries can teach Libra how to speak up and stand out of the shadows, while Libra teaches Aries how to calm down and return to their center of inner balance.

Taurus + Scorpio:

This is a fixed combination and they both like to focus on money and the valuable things in life. With earthly delights, Taurus can teach Scorpio how to get grounded and enjoy the simple pleasures in life. Whereas Scorpio likes to teach Taurus how to get out of their comfort zone and explore the unknown!

Gemini + Sagittarius:

Two mutable signs that are flexible in their nature. Gemini urges Sagittarius to focus on the details, while Sag teaches Gemini to look at the big picture instead. These two will balance each other out and have a lasting relationship where both small and big plans can be made!

Cancer + Capricorn:

The other cardinal combination on the block, these two are like the yin to the yang. Cancer retreats inside but is coaxed out to the practical world by Capricorn. Whereas, Capricorn can learn how to go inwards and feel their emotions more, in this dynamic combination.

Leo + Aquarius:

Another fixed relationship, these signs like to be seen and like to be innovative. They both know how to party and will support each other in their wild ideas. The trouble will be learning who's turn it is in the spotlight, but with open communication, these two will work it out just fine.

Virgo + Pisces:

Grounded Virgo meets dreamy Pisces, to create a cardinal combination. The common ground is that they both love to help others, but Virgo will be teaching Pisces to return to earth and implement a little common sense. In this time, Pisces will be bringing Virgo out of the analytical mind and into the dream state. Both important aspects to embody!

As you see from these polarities, there are lessons to be learnt in each relationship, but dating the opposite sign is a sure way to make the differences obvious and the lessons big. Once you can make it past the nitty gritty parts, you may even begin to enjoy the aspects of your partner that are so different to your own. That's how we all grow and evolve, right?

SHOULD THE SAME STAR SIGNS DATE?

We have looked at opposites dating, but how about the same signs getting together? In some cases, this can be dreamy and other cases, it can feel more like a nightmare!

Dating someone similar to you can have its benefits, as that person already knows what you like and what your pet peeves are. This can help you get along nicely for a while, but the downside is that people often get bored of the same thing after a while!

Are you seeking an easy and compatible partner or are you looking to challenge your boundaries and grow as a person? Getting to know someone on the first few dates is often the most magical part of relationships, and this can get boring if every answer is always the same as yours! In other cases, people may like this similarity, for it already shows commitment and stability.

Here are the signs that work best together:

Taurus meeting a Taurus is like a home coming. Immediately, they can get comfortable and relax knowing that their partner is in it for the long run, just as much as they themselves are. Although their stubborn minds can clash, they often have similar points of views on most topics anyway.

Two Leos also make a great pair! Although the lion gets a wrap for being boastful, most of them are actually big hearted and caring for their partners. This makes a fun and loving partnership that can feast on life together.

Libras are another great pair. They are about balance after all, so they know how to maintain balance within a relationship. They are also aware of their tendency to create harmony rather than speaking their truth, but this is something that they can work on together.

Sagittarius couples are another fiery combination. You will see them travelling the world, laughing and joyful as they challenge themselves to step out of their comfort zones. They are an easy going pair as they both respect each other's need for independence.

Two Virgos can also be a power couple. Organised, grounded in their bodies and excited to take action to achieve their goals, this combination is ready for anything.

Aquarius and Aquarius can also be a blessing. Usually, these two signs can go together well if they put their minds together on with their humanitarian hearts. Helping to upgrade and improve the world is something they will definitely bond over!

Capricorns are loyal and caring, and a pair of two Capricorns will find a steady bond and stable relationship. This will be a pairing to count upon.

Scorpio pairs are also potent. They understand the mysterious sides of their partners and they use these depths to explore the universe and uncover more truths. This pairing will be intense but deeply passionate and loving.

Signs that shouldn't date their own sign:

Aries, we are looking at you. Two hot headed partners can cause surface conflict regularly and that's just exhausting. Also, you're bound to go off in two different directions, because no Aries likes to be a follower.

Geminis are also a little complicated as they are airy and flighty, and they usually need a solid partner to keep them grounded.

Cancers are too emotional with other cancers, and this combination might lead to an emotional black hole that neither can climb out of.

Pisces partners are also an insightful combination as long as one of them stays connected to what's happening down here on earth. But the downside is that when both are up in the clouds, they might find that they are dating the dream version of their partner rather than seeing them for who they actually are. This can make the relationship easy at first, but perhaps more challenging in the long run when reality comes crashing down

CONCLUSION

As you know, dating is complex, and there is a lot to know about a person before choosing them as your future partner. But keep in mind that some of these combinations will help you to choose a partner that will be compatible with your sign. This is the general rule of thumb, but there are always exceptions. So don't throw your partner out because they have the wrong sign. Look to their Venus or Mars instead. Or, even better, see an astrologist to do an in-depth understanding.

Now as you move forwards in your relationships, you can also use this information to better understand what you and your partner seek in your relationship. When you are clear on what you both need, it can be easier to live in a harmonious and loving way, where both of you grow through challenges to come out the other side wiser and more open-hearted.

Conclusion

The stars have been burning brightly for thousands of years and will continue to illuminate the sky for many more lifetimes to come. This book was a soft introduction to the big wide world of astrology, and I hope that you have landed gently, with the energy and passion to go forth and pursue more books and courses about the stars and planets. I encourage you to closely examine your chart, start reading astrological insights for others, and to continue to quench your thirst for more astrological knowledge. There is so much out there, and the best way to absorb the lessons is the apply the knowledge to yourself. As you have learnt, astrology is a practical tool – and now you have the wisdom to implement it into your daily life.

Take some time after reading this book to set both short-term and long-term goals based on your personality traits and what you think your 'sign' would like to explore.

Perhaps your moon in Taurus is daring you to try a new cooking course. Maybe your sun in Sagittarius is inviting you to book your next holiday. Your Leo rising could be challenging you to find new ways of creating abundance or your Pisces sun wants you to search for more astrologer friends!

Think about the aspects in your chart that surprised you and consider the ways in which you can invite that area of yourself to be more present in your life. This may mean getting out of your comfort zone (fine if you're an Aries Sun, but not so fine if you're a Taurus), but you could be surprised at the fun you might have! Maybe you will even find a part of yourself that you didn't know existed.

Throughout this book, you may have realised that you are an important part of the universe. You are connected to the sky above and the earth below. You are here to accomplish a unique purpose, and the stars are ready to tell your story.

By now, you should have stopped blaming yourself for being different. You were never made to fit in with others, you were made to be your own exclusive part of the universal puzzle. The exact placement of the planets, signs and houses in your chart are different to every other person on this earth. So, embrace your inimitable qualities, spread your magic and know that it is never too late to become the astrologist that you may have dreamed of being.

You have immersed yourself in both your inner and outer worlds. You have explored aspects of yourself that most people don't dare to look at. You have challenged your notion of what is right and wrong for you in society. You now know that you are being supported by the stars to reach your goals and accomplish your dreams.

If you did enjoy this book, then it is time to spread this knowledge together. You can help to make astrology more wide-spread and accepted throughout our world by writing a review about this book! The more you spread the word, the more people can rise to a new level of self-understanding and self-acceptance. So if you can, please take one minute to leave a review and include your sun sign, moon sign and rising sign as your signature so we know what mystics are being drawn to the world of the stars!

About the Author

Ashley was born in a little village just outside of Ojai, California.

She grew up surrounded by open-minded people, like her mother, who worked as a herbalist for the majority of her life. Ever since she was little, Ashley watched her mom skilfully investigate the planetary connections of plants, and understand the alignment of plants with individuals' birth charts. She learnt from a young age the powerful role that astrology and botany have in healing others.

Ashley has always been fascinated by her mother's work, joining many 'witch circles' with her mom to learn more about the ancient art.

After finishing her 5-year long degree in Naturopathy, Ashley attended many courses about astrology, human design, crystals, and shamanic ways of healing.

She is deeply passionate about helping others achieve clarity about themselves. She often says that astrology has supported her throughout her entire life and been the driving force which helped her surpass her struggles and reach peace.

Since achieving her own inner peace, Ashley has used her knowledge and experiences to help many others across the globe to understand their own personalities, to find their perfect lovers, to attract financial abundance and to, essentially, find their *true* purpose!

In this book, Ashley has aimed to give readers just a small taste of the vast amount of knowledge she has on the subject of Astrology. She wants this book to be the start of an ongoing awakening for all and truly hopes that her writing speaks to the souls of all who read it!

Also Published by Master Plan Focus

MOTIVATIONAL STORY

BOOK 1

Inspirational Short Stories from Everyday Life, with a Thought Provoking Anecdote

ANDY HAYNES

If you're looking for **a book that will inspire and motivate you**, then look no further than this **Motivational Story Book**. It has **50 inspiring short stories** from everyday life, each with a **thought-provoking anecdote**.

These are the kind of stories that **stay with you** long after you've finished reading them. They'll make you laugh, cry, and think deeply about what it means to be human.

Each one is different, but they all have the same goal - to help you see the possibilities for your own life and to find the motivation to **pursue your dreams**.

Whether you're facing a challenge in your personal life or professional life, this book will give you **the motivation you need** to overcome any obstacle. So don't hesitate, buy the Motivational Story Book today and start living your best life!